Contemporary Issues in Technology Education

Technology education is a developing field, new issues keep arising and timely, relevant research is continually being conducted. The aim of this series is to draw on the latest research to focus on contemporary issues, create debate and push the boundaries in order to expand the field of technology education and explore new paradigms. Maybe more than any other subject, technology education has strong links with other learning areas, including the humanities and the sciences, and exploring these boundaries and the gaps between them will be a focus of this series. Much of the literature from other disciplines has applicability to technology education, and harnessing this diversity of research and ideas with a focus on technology will strengthen the field. For more information, or to submit a proposal, please email Melissa James: melissa.james@springer.com

More information about this series at http://www.springer.com/series/13336

Gabriel J. Costello

The Teaching of Design and Innovation

Principles and Practices

Springer

Gabriel J. Costello
Galway-Mayo Institute of Technology
Galway, Ireland

ISSN 2510-0327 ISSN 2510-0335 (electronic)
Contemporary Issues in Technology Education
ISBN 978-3-030-41379-8 ISBN 978-3-030-41380-4 (eBook)
https://doi.org/10.1007/978-3-030-41380-4

This Springer imprint is published by the registered company Springer Nature Switzerland AG
The registered company address is: Gewerbestrasse 11, 6330 Cham, Switzerland

To Mary with love

Preface

My background and interest in design and innovation is grounded in 20 years of industrial experience in product development and new product introduction. The subsequent teaching of the subject in a third-level institute has been an interesting and challenging voyage through a largely unchartered territory. Design and innovation is always a journey into the unknown, and its teaching must accommodate this scenario. My aim is to distil my experience in order to provide some guiding principles and practices for others undertaking this journey.

Purpose and Motivation for This Book

There is a growing and varied body of publications and web resources available on the topic of design and innovation. However, there is a lack of studies on how to teach the subject. The objectives and contributions of this book are as follows:

- To support the teaching of design and innovation in a wide range of educational institutes.
- To provide a philosophical and theoretical underpinning to the education of design and innovation.
- To provide case studies of teaching design and innovation.
- To propose the concept of simulation-action learning (SAL) in order to extend the pedagogical approach.
- To engage in the debate on contemporary issues in technological education.

Who Will Benefit from This Book?

There are three cohorts for which this book is intended.

1. Capstone undergraduate and postgraduate programmes.
    - To provide a framework for the teaching of design and innovation modules across all engineering disciplines.
    - To inform programmes specialising in industrial design or related disciplines.
    - To support programmes in information and communications technology (ICT) and information systems (IS) that include modules on business process design or similar.
    - To enable the introduction of "enterprise" modules in masters and doctoral programmes that can instil the motivation and skills required for the application of research as products or services in commercial or not-for-profit situations.
    - To assist the delivery of innovation and entrepreneurial modules in science and business schools.
2. Continuing professional development programmes.
    - To enable the development of higher-order design and innovation capabilities in industry and public service organisations to meet contemporary needs.
    - To provide a framework for the development of cross-disciplinary teams required to address complex organisational and so-called societal wicked problems.
    - To facilitate organisations moving from being product-centric to being service-centric.
3. Research programmes.
    - To provide a resource for future research and debate on the teaching of design and innovation and to contribute to the development of the broader area of design and innovation research.

Galway, Ireland Gabriel J. Costello

Acknowledgements

This book would not have been possible without the friendship and support of my colleagues in the Galway-Mayo Institute of Technology and industry. I am also grateful to my students whose innovative designs and thought-provoking feedback provided a basis for the work. I am indebted to the anonymous reviewer for the excellent advice and constructive suggestions. Thanks also to my Springer Editor, Helen van der Stelt, for her kind assistance in the preparation of the manuscript. I am especially grateful to my wife Mary, son Mark and daughter-in-law Susannah, grandchildren Orla and Róisín, son Colm, daughter Maria and son-in-law Marcus for the love, encouragement and happiness they continually bring to my life.

Galway, Ireland Gabriel J. Costello

Acknowledgements

[illegible]

Contents

Part II Practices

About the Author

Gabriel J. Costello is a Lecturer in Engineering at the Galway-Mayo Institute of Technology since 2001. Prior to this, he worked for 20 years in the telecommunications industry where he held engineering, supply chain and product management positions. In 2010, he completed his PhD in Management Information Systems at the J.E. Cairnes School of Business & Economics, National University of Ireland, Galway, in the area of information systems innovation. His 2004 Masters by Research (MEng) examined the diffusion of energy management practices in Ireland's SME sector and was supported by the Sustainable Energy Authority of Ireland (SEAI).

His research interests include the management, theory, philosophy and teaching of design and innovation and the sociotechnical challenges of the transition to sustainable energy communities.

Chapter 1
Introduction

This book is about design and innovation – what it is and how to teach it. The blending of design and innovation is having an increasing impact not only on the world of product and services but on a wide variety of disciplines such as information and communications technology (ICT), business, education and medicine. Additionally, its application is relevant to public sector bodies and the not-for-profit sector, which have been largely overlooked in the literature. While there are numerous books on the topics of design and innovation, there is a dearth of literature on how to teach it to undergraduate and graduate cohorts and to guide practitioners on how to embed the approach in vocational and professional development programmes. Meredith Davis (2017) expresses her surprise that there are so few books on teaching design and innovation given the significant growth of interest in both workplace and academia. Furthermore, she contends that most design teachers have little preparation for their role. She is also concerned by the perception of many practitioners that formal education has little to offer the discipline. This book will address this gap by outlining foundational principles behind the teaching of design and innovation and by offering a practical process for implementing the pedagogy in academic institutions and outside academia in the context of continuing professional development (CPD). The book describes my journey in formulating underlying principles to inform the practices that I use to teach the subject. My hope is that it can help others on their own unique journey through the challenging but fascinating terrain that is design and innovation.

Design and innovation is as old as civilisation, but its formulation as a recognised process is of more recent origin. For example, Alfred North Whitehead, the British mathematician and philosopher, once remarked that "the greatest invention of the nineteenth century was the invention of the method of invention". The modern formulation of the concept of design is often located in the seminal publication by Herbert Simon of *The Sciences of the Artificial* (Simon, 1996), while many scholars trace the introduction of the concept of innovation into the realm of economic and social change to Joseph Schumpeter's influential work: *The Theory of Economic Development* (Schumpeter, 1934). According to Nigel Cross, design and innovation

G. J. Costello, *The Teaching of Design and Innovation*, Contemporary Issues in Technology Education, https://doi.org/10.1007/978-3-030-41380-4_1

is a journey to explore and discover something new; but it is a complex journey. Furthermore he points out that coping with uncertainty is a key aspect of ability in this area, and he contends that if you prefer the certainty of structured well-defined problems, you will never appreciate the delight of being a designer (Cross, 2011). The literature on design and innovation is wide-ranging and voluminous, and a fitting appraisal is outside the scope of this work. However, I will now provide a brief introduction to concepts relevant to the theme of its teaching.

1.1 The Case of Design

Design thinking is having an increasing influence in the areas of teaching and research. The approach has its academic origins in the Stanford School of Mechanical Engineering in the 1970s. Now it integrates business, law, medicine, the social sciences and humanities into more traditional engineering and product design education. The design thinking method proposed by the *Hasso Plattner Institute of Design at Stanford* originally consisted of five stages: empathise, define (the problem), ideate, prototype and test. However, the pedagogy has evolved to a focus on enabling students in eight core design abilities as shown in Fig. 1.1 (d.school, 2019).

Design thinking has been adapted to various situations such as when it is applied to an educational setting (Dym, Agogino, Eris, Frey, & Leifer, 2005). Furthermore, the Stanford approach is having increasing influence in higher educational institutes around the world. For example, the *Sugar Design Thinking Global HEI Network* managed by the University of St. Gallen is an example of its growing global

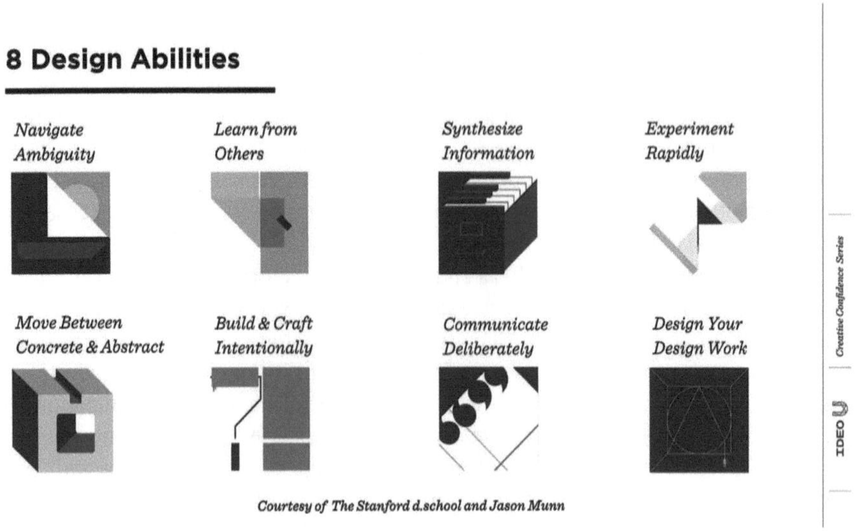

Fig. 1.1 d.school 8 core abilities

educational influence (SUGAR, 2019). The advantage of design thinking is well summarised by the Porto Design Factory based in the Instituto Politécnico do Porto (P.Porto, 2019).

> Unlike most other engineering design courses, design thinking requires the students to get out into the field and interview users to understand people's values as well as thoroughly benchmarking existing products and technologies. The iterative nature assures that teams are not stuck on one idea for too long and that ideas are being continuously validated with end users through rapid prototyping and testing. (ME310Porto, 2019)

However, my experience of contemporary global complex product design projects compels me to offer a word of caution. Design thinking provides a very exciting and ground-breaking "front end" to the design process, but its presentation as a panacea by some commentators can damage its credibility. There is a long way to go from conceptualising and prototyping a design to its realisation as products, services, solutions, systems, architectures, spaces or experiences. Furthermore, the teaching of how to develop a design into a feasible venture, whether for-profit or not-for-profit, is largely missing in the literature, and I intend to address this. Gordon Murray, renowned racing car designer, cautions that design consists of 1% inspiration and 99% perspiration, which explains why many enthusiastic people loose heart and abandon the process. The enhancement brought about by design thinking is that it transfers a significant proportion of the perspiration to the "front end". It is in the context of the realisation of design thinking *through* innovation that I will now briefly introduce the latter topic. I realise that there are several other "design" methodologies such as user experience (UX) design and universal design that are currently in use. However, similar principles apply, and I have used design thinking as an exemplar given its growing global influence in education.

1.2 The Case of Innovation

Innovation is now a major focus for organisations, regions and economies, and the subject is increasingly seen as being crucial not only to success but to survival. According to Brynjolfsson and Saunders (2009, p. ix), the fundamentals of the world economy indicate that there will be a continuation of innovation "through the booms and busts of the financial markets and of business investments" (p. ix). As Becerra (2009) points out, innovation "is the oil of our economic system that keeps it continually running in search for greater value to customers" (p. 123). Schumpeter's influential work on *The Theory of Economic Development* classified innovation into five categories: new products (or goods), new methods of production (or processes), new sources of supply (or half-manufactured goods), the exploitation of new markets and new ways to organise business. According to Swedberg (1991), in Schumpeter's original schema, innovation is accomplished by "entrepreneurs" who developed new combinations of existing resources. However, in his later works, he came to regard the large corporation as the innovative engine driving the development of leading economies (Lazonick, 2005). Fagerberg (2005, p. 4) makes the fundamental distinction between invention and innovation where the former is regarded as the "first occurrence" while the latter is the "first attempt to carry it out into practice". This is in line with Van de Ven's (1986, p. 604) assertion that "an invention or creative idea does not become an innovation until it is implemented or institutionalized". To put it simply, there is no innovation until someone is using it. Zaltman, Duncan, and Holbek's (1973) contingency theory of innovation predicts that the effect of structural variables will be conditional on the two main subdivisions of the innovation process: the initiation stage and the implementation stage. According to these authors, the most important contribution by James Wilson as part of his theoretical work on innovation in the 1960s was the identification of the innovation dilemma which organisations face during the process of innovation. Wilson (1966) had concluded that it is easier to initiate than implement innovations since organisations can quickly increase capacity to generate new proposals but find it much more difficult to match this capacity in the ratification phase.

1.3 Design-Driven Innovation

Verganti (2009) has proposed new agenda for design and innovation studies by unveiling a novel strategy which he calls design-driven innovation. This perspective, he argues, goes beyond the focus on radical innovation (pushed by technology) or incremental innovation (through market pull). He concludes that a significant finding from the literature on innovation in the past decades is that "people do not buy products but meanings". He contends that people use things not just for utilitarian reasons but for "profound emotional, psychological and sociocultural reasons as well" (ibid., p. 4). Verganti uses the case study of an Italian lighting company, Artemide, who have adopted the design-driven strategy by designing lamps that not only give you light but have a different and unexpected meaning: the light makes you feel better. Firms such as Artemide step back from users and take a broader perspective. These companies go beyond developing technologies, products and services to envisioning how their innovations could change life contexts for the better. They are not simply following trends but making propositions that will modify the context in which they are used. Another example he uses is Swatch, whose innovations changed the context of a watch from being a technological instrument (albeit with a vast range of innovative features), to a fashion accessory matching clothing styles and even the seasons. Swatch developed a new meaning for the humble timepiece. Importantly he proposed that meanings, like technologies, can be subject to a research and development (R&D) process. Verganti and Dell'Era (2014, p. 140) propose that design is related to the innovation of the meanings of products and services and concerns itself with "why" people use things and not just with their functionality and performance (i.e. the "what" and "how"). Furthermore, they define the process as follows (ibid., p. 146):

> Design-driven innovation: where innovation starts from the comprehension of subtle and unspoken dynamics in socio-cultural models and results in proposing radically new meanings and languages that often implies a change in socio-cultural regimes.

The exploration of these meanings is the result of an interpretive process, and successful adopters explore the context in which people's lives are evolving in sociocultural terms (how the reason why people buy things is changing) and in terms of how technology is shaping the context of a person's world. In separate work, Lester and Piore (2004) propose *interpretive spaces* where innovation is viewed in terms of developing new language communities rather than the conventional approach of rational decision-making. Companies that adopt this view are interpreters, which involves three stages: listening, interpreting and addressing and where innovation through design becomes a hermeneutic process (Verganti & Dell'Era, 2014, p. 156). This interpretive theme will be examined in relation to the empirical cases presented in Chap. 6. The design-innovation literature contains some philosophical and conceptual challenges but offers a thought-provoking perspective on the complexity of the contemporary world of design and innovation. This book aims to address these issues particularly in the opening chapters.

According to Katz (2004), the main issues facing innovation managers are not the technical area but in overseeing the complex interplay and motivation of the people involved. The challenge includes the ability to "communicate, understand, inspire and guide" and to integrate people within broader organisational areas such as finance and marketing. In addition, the emerging not-for-profit sector is an important source of societal innovation despite its paucity in the mainstream literature. An objective of this work is to support the growing focus on the human aspects of design and innovation, in the principles that are adopted and the practices that are proposed. In summary, there is almost universal agreement that the synthesis of design and innovation is a complex phenomenon to understand and manage (Allen, 2004; Eppinger, 2001; Goffin & Mitchell, 2005; Katz, 2004; OECD, 2005; Poole & Van de Ven, 2000; Rothwell, 1994). It is in this context that the book has been developed, to emphasise the practical application of both philosophy and theory using team-based pedagogy embedded in real-world project scenarios (Costello, 2019).

1.4 Teaching Design and Innovation

A starting point to the discussion on the teaching of design and innovation is to understand that there is a significant difference between the scientific method and the design method. Gregory (1966a, p. 6) argued that designers could not hope simply to *copy* the scientist's method since designers and scientists have radically different goals. "The scientific method is a pattern of problem-solving behaviour employed in finding out the nature of what exists, whereas the design method is a pattern of behaviour employed in inventing things of value which do not yet exist, Science is analytic, design is constructive". Gregory's work was the fruit of a seminal conference on the nature of "the design method" organised by the Design and Innovation Group in the University of Aston in Birmingham, England (S. A. Gregory, 1966b). The fact that the book has been recently republished is a testament to the enduring influence of this body of work (S. A. Gregory, 1966c). Cross, Naughton and Walker (1981, p. 195) build on the conference debate and argue that "attempts to model the design method on the scientific method seem misplaced" especially in view of what they regard as the contemporary epistemological chaos surrounding the concept of the "scientific method" (ibid., p. 200). They propose that design is viewed as a technological activity. However, they point out that the term "technology" must not be interpreted in the narrow sense of machinery, applied science and engineering. Their definition is much wider: "Technology is the application of scientific and other organised knowledge to practical tasks by social systems involving people and machines" (ibid., p. 198). They contend that design as practised is strongly directed towards "practical tasks, solutions or actions" and involves a variety of kinds of knowledge that includes craft, skill and know-how. These points, I believe, resonate with Aristotle's concept of phronesis examined in Chap. 2. In related work, Gedenryd (1998) in his thesis, *How Designers Work: Making Sense of Authentic Cognitive Activity,* argued that classic methods of reasoning in problem-solving are inappropriate in design. Herbert Simon (1996, p. 5) sums it up in his influential book as follows: "natural sciences concern itself solely with how things are", while "design on the other hand is concerned with how things ought to be, with devising artifacts to attain goals". My deduction from these augments is that the teaching of design and innovation cannot be carried out in the same way as the teaching of science. Addressing this conclusion and proposing an alternative path are one of the motivations for this book.

King's (1993) paper *From Sage on the Stage to Guide on the Side* sums up this book's approach to delivering content and creating the educational atmosphere for teaching design and innovation. She concluded that the transmittal model of education, where the teacher is the central figure "sage on the stage", was outdated. The approach was not fit for purpose for teaching in the twenty-first century when "individuals will be expected to think for themselves, pose and solve complex problems and generally produce knowledge rather than reproduce it" (ibid., p. 30). A new approach was required where students were now at the heart of the educational process, where they would take active part in thinking about and discussing ideas with the view to sense-making. Here the teacher becomes a facilitator of learning and takes the role of "guide on the side". However, it is important to realise that King still emphasises that the teacher is responsible for the course material but now presents it in such a way that the students engage and interact with the content. The teacher in this new approach "orchestrates the context, provides resources and poses questions to stimulate students to think up their own answers" (ibid., p. 30). Interestingly, the philosopher, Erasmus of Rotterdam, when describing the role of a teacher, pointed to the etymology of the word "education" coming from the Latin *educere* to "lead out". Consequently, the role of the design and innovation teacher is to provide the environment to draw out from students, things which did not previously exist.

Leifer and Meinel (2016) have recently proposed a "Manifesto" which argues that design thinking is mature enough to become foundational for engineering education alongside scientific fields such as physics and chemistry. This book is an attempt to help support this goal by providing teachers with a pedagogical framework to implement design and innovation in an educational context. It is an approach to the topic which is not confined to the classroom but engages with real-world practice scenarios. Such approach will require ambidextrous design skills and competences (Tushman & O'Reilly, 2004). As a result, the cultivation of Aristotle's concept of phronesis – doing the right thing, in the right place, at the right time, in the right way – is examined in Chap. 2.

In summary this book is not about *the teaching of the process of design and innovation* which has a wide-ranging body of literature (Cooper & Kleinschmidt, 1993; Cross, 2000; Curedale, 2013; Ulrich, Eppinger, & Yang, 2020). Rather it is concerned with the *process of teaching design and innovation* which has a paucity of publications, and so it is primarily intended as a resource for teachers and instructors.

1.5 Learning for Design and Innovation

A report prepared for British Department of Education and Science in 1979 supports the claim that design represents a distinct way of knowing. The researchers argued that the two dominant views of knowledge, i.e. science and humanities (or arts), required the addition of a third domain, which studied the "conception and realisation of new things". The thesis of this book is that an alternate way of knowing requires a different way of teaching and learning. The pedagogical approach to learning design and innovation draws on Laurillard's *Conversational Framework* (Laurillard, 2012a). Her model describes the conditions necessary for learning to take place and consists of six categories: acquisition, inquiry, practice, production, discussion and collaboration as shown in Fig. 1.2. Furthermore, she concludes that the widespread embracing of technologies by both teachers and students has profound implications for teaching and learning. To meet these challenges, she proposes that teaching should be regarded as a design science which builds principles rather than theories. Using this concept of design is particularly relevant to the theme of this book, which focuses on the principles and practices required for teaching design and innovation. It is important to note that Laurillard (2012a p.1) acknowledges that teaching as design science uses and contributes to theoretical science. Hence, I will explore the topic of theory in Chap. 3.

Table 1.1 shows a rage of methods and technologies proposed by Laurillard. The table outlines each of the six learning types and provides examples of conventional and digital approaches that serve each type.

Laurillard's table is a convenient checklist to examine the scope of a teaching and learning programme and will be referred to in Chap. 7 which outlines a pedagogical process for design and innovation.

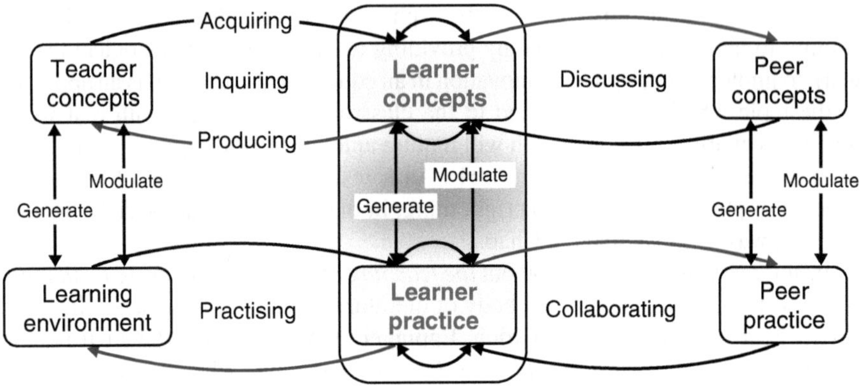

Fig. 1.2 Laurillard's six learning types – slide from (Laurillard, 2012b)

Table 1.1 Types of learning with conventional and digital technologies (Laurillard, 2012a, p. 96)

Learning through	Conventional technology	Digital technology
Acquisition	Reading books, papers Listening to teacher presentations Face-to-face, lectures Watching demonstrations Master classes	Reading multimedia, websites, digital documents and resources Listening to podcasts, webcasts Watching animations, videos
Inquiry	Using text-based study guides Analysing the ideas and information in a range of materials and resources Using conventional methods to collect and analyse data Comparing texts, searching and evaluating information and ideas	Using online advice and guidance Analysing the ideas and information in a range of digital resources Using digital tools to collect and analyse data Comparing digital texts Using digital tools for searching and evaluating information and ideas
Practice	Practising exercises; doing Practice-based projects, labs, field trips, face-to-face role-play activities	Using models, simulations, microworlds, virtual labs and field trips Online role-play activities
Production	Producing articulations using statements, essays, reports, accounts, designs, performances, artefacts, animations, models, videos	Producing and storing digital documents, representations of designs, performances, artefacts, animations, models, resources, slideshows, photos, videos, blogs, e-portfolios
Discussion	Tutorials, seminars, email Discussions, discussion groups, online discussion forums, class discussions Blog comments	Online tutorials, seminars Email discussions, discussion groups, discussion forums Web-conferencing tools, synchronous and asynchronous
Collaboration	Small group project Discussing others' outputs Building joint output	Small group project Using online forums, wikis, chat rooms, etc. for discussing others' outputs Building a joint digital output

1.6 Team-Based Learning

Teams have long been described as the fundamental learning units in the modern organisation (Mike Pedler, Burgoyne, & Boydell, 1991) and are being used effectively in areas related to design and innovation such as product development, process-centred organisations and project management. Teamwork is integral to design and innovation which is rarely carried out successfully by the romanticised lone inventor. This necessitates the assembly of heterogeneous team members. For example, the importance of a cross-functional design team in this modern multidimensional environment is stressed by Cross (2011) who describes design as "as a shared social process".

The origin of team-based learning (TBL) is usually attributed to Larry Michaelsen working in the faculty of the University of Oklahoma during the 1970s. Michaelsen differentiated between the power of *teamwork,* over and above *group work,* and proposed TBL as an instructional strategy rather than merely an instructional technique (Michaelsen, Knight, & Fink, 2004). According to Fink (2004), the new approach was due to a number of factors, such as appeals by students for a learning experience that was more than just an "information dump" and feedback from employers that in addition to content knowledge, human interaction competences and problem-solving ability was required from graduates. Furthermore, Fink (ibid., p. 4) contends that well-implemented TBL results in four kinds of transformations:

1. It transforms "small groups" into "teams".
2. It transforms a technique into a strategy.
3. It transforms the quality of student learning.
4. For many teachers, it transforms (or restores) the joy of teaching.

In relation to point one, he makes the important distinction between small groups and teams. The latter must be properly formed, remain together long enough to become cohesive teams and be repeatedly given challenging tasks with prompt and clear feedback (ibid., p. 8). In relation to point two, he explains that TBL is "a set of learning activities in a particular sequence, not just an individual activity or teaching technique that can be plugged in more or less wherever" (ibid., p. 9). Consequently, TBL requires semester-long instructional strategies that have a clear set of linked activities to deepen student learning. This will enable the development of a united team (recommended to have five to seven members) rather than just a short-term assembly of a group. This approach will inform the development of the instructional processes outlined in Chap. 7 of this book. Michaelsen (2004) points out that the main aim of small group cooperative teamwork is to *learn concepts*, while the aim of team-based learning is the *application of learned concepts* where the teacher becomes the manager of the instructional process to achieve this. Teams should be diverse which, he cautions, usually results in short-term issues. However, once resolved by the team members themselves, this diversity becomes a major asset during the extended TBL timeframe. Michaelsen (2004, p. 28) proposes four essential principles of team-based learning which I will list here and use later in Chaps. 6 and 7 (Table 1.2).

Table 1.2 Michaelsen's principles of team-based learning

1.	Groups must be properly formed and managed
2.	Students must be made accountable for their individual and group work
3.	Group assessments must promote both learning and team development
4.	Students must have frequent and timely performance feedback

In Chap. 6, I will present the case of collaboration between engineering students and business students, with its advantages and challenges, in a higher educational environment. Furthermore, I will use Michaelson's principles to evaluate my proposed instructional framework in order to facilitate its implementation in schools that have limited resources.

Now I will briefly outline the themes of the book and provide a short introduction to the chapters.

1.7 Layout of the Book

Part I: Chaps. 2, 3, 4 and 5 will outline the *principles* of teaching design and innovation.

Part II: Chaps. 6, 7 and 8 will outline the *practice*s (and challenges) of teaching design and innovation.

Chapter 2 will introduce the topic of philosophy and its importance for design. The educational philosophy of a programme is an important statement of the underlying approach to teaching and learning (O'Neill, 2015). Consequently, I will propose Aristotle's principle of *phronesis* as suitable foundation for the teaching of design and innovation. The concept of phronesis (*phronēsis*) has been described as practical wisdom with connotations of intelligence and soundness of judgement.

Chapter 3 will introduce the principle of responsible innovation (RI) and examine its implications for the teaching of design and innovation. Empathy is an early and important step in the design and innovation process. This chapter will introduce Edith Stein, the philosopher of empathy, and discuss the implications of her work for teaching design and innovation. Furthermore, it will ground this examination in the ethical lens of Emmanuel Levinas.

Chapter 4 will examine the principle of theory and outline why it is of value when conceptualising the topic of teaching design and innovation. One of the most challenging tasks in research and teaching is to describe *what a theory is* and *why it is important*. This chapter will outline the development of a metatheory of design and innovation that expands the traditional view which holds that resources and knowledge are the most important attributes of an organisation. This metatheory will inform the development of the pedagogical framework in Chap. 7.

Chapter 5 will argue that the teaching of design and innovation requires a robust methodical approach informed by the principle of *reflection*. The methodology will be presented in the context of action learning (AL) and action research (AR)

grounded in Donald Schön's work on the reflective practitioner and in Edmund Husserl's phenomenology of reflection. The chapter will also provide a novel pedagogical suggestion on embedding the habit of reflection-in-action in teaching practice through regular use of agile development methodologies.

Chapter 6 will describe two undergraduate case studies that aimed to instil design and innovation competences in students of both engineering and business disciplines. The cases involved student teams working with incubation centre start-ups and multi-national subsidiaries. One of the aims of this book is to provide a resource for the continuing professional development (CPD) of people in industry, the service sectors and the public sector. Consequently, a practitioner-based case study will be presented that outlines the benefit of academic-practitioner dialogue and the creation of interpretive spaces for the development of innovative practice. Furthermore, this case is presented in the context of research-informed teaching that supports the framework developed in Chap. 7.

Chapter 7 will propose a pedagogical framework, simulation-action learning (SAL), as an approach to the practice of teaching design and innovation. The methodology was derived from the real-world case studies in Chap. 6, honed by the philosophical and reflective approaches in Part I of the book. The framework is demonstrated as consisting of seven phases, and recommendations for each phase are proposed. However, the phases will be presented as placeholders and patterns and not as prescriptive tasks. This approach is justified due to the iterative nature of design and innovation, together with the reality that each phase will be revisited, sometimes more than once, in the design journey. Furthermore, the recommendations encompassing each phase are offered as flexible suggestions for teachers operating in their own unique environment.

Chapter 8 will describe some contemporary issues facing the teaching of design and innovation with a special focus on topics such as the debate on artificial intelligence (AI); the technological explosion generated by digitalisation; the implications of Alasdair MacIntyre's *After Virtue* for teaching; and the effects of the dominant influence of positivism on scholarship. The focus of this chapter is to respond to the challenges set by the editors of this series, to stimulate debate and push boundaries in technology education. Each vignette will be linked to the topics presented in the preceding chapters.

The principles are examined in detail in Part I, but it should be remembered that the function of the principles is to serve practice. The process of design and innovation is not linear, and there are many iterations and revisiting of the stages throughout its meandering journey.

Consequently, a teacher can begin with Chap. 1 and, in the spirit of learning by doing, move immediately to Chaps. 6 and 7 and then engage with the other chapters during their own learning and reflection process. My aim is not be prescriptive but descriptive, in the hope that this book will be of assistance to others who have an interest in this area.

References

Allen, T. J. (2004). Communication networks in R&D laboratories. In R. Katz (Ed.), *The human side of managing technological innovation: A collection of readings* (2nd ed.). Oxford, UK: Oxford University Press.

Becerra, M. (2009). *Theory of the firm for strategic management: Economic value analysis*. Cambridge, UK: Cambridge University Press.

Brynjolfsson, E., & Saunders, A. (2009). *Wired for innovation: How information technology is reshaping the economy*. Cambridge, UK: MIT Press.

Cooper, R. G., & Kleinschmidt, E. J. (1993). Stage gate Systems for new Product Success. *Marketing Management, 1*(4), 20–29.

Costello, G. J. (2019). The philosophy of innovation in management education: A study utilising Aristotle's concept of Phronesis. *Philosophy of Management, 18*(3), 215–230. 10.1007/s40926-018-00104-7.

Cross, N. (2000). *Engineering design methods: Strategies for product design*. Chichester, UK: Wiley.

Cross, N. (2011). *Design thinking: Understanding how designers think and work*. Oxford, UK: Berg Publishers.

Cross, N., Naughton, J., & Walker, D. (1981). Design method and scientific method. *Design Studies, 2*(4), 195–201.

Curedale, R. (2013). *Design thinking: Process and methods manual*. Topanga, CA: Design Community College.

d.school. (2019). *Hasso-Plattner Institute of Design at Stanford* (d.school). https://dschool.stanford.edu/. April 2019.

Davis, M. (2017). *Teaching design: A guide to curriculum and pedagogy for college design faculty and teachers who use design in their classrooms*. New York: Allworth.

Dym, C. L., Agogino, A. M., Eris, O., Frey, D. D., & Leifer, L. J. (2005). Engineering design thinking, teaching, and learning. *Journal of Engineering Education, 94*(1), 103–120.

Eppinger, S. D. (2001). Innovation at the speed of information. *Harvard Business Review, 79*(1), 149–158.

Fagerberg, J. (2005). Innovation: A guide to the literature. In J. Fagerberg, D. Mowery, & R. R. Nelson (Eds.), *The Oxford handbook of innovation* (pp. 1–26). Oxford, UK: Oxford University Press.

Fink, L. D. (2004). Beyond small groups: Harnessing the extraordinary power of learning teams. In L. K. Michaelsen, A. B. Knight, & L. D. Fink (Eds.), *Team-based learning: A transformative use of small groups in college teaching*. Sterling, VA: Stylus Publishing.

Gedenryd, H. (1998). *PhD Thesis: How designers work – Making sense of authentic cognitive activities*. Lund University Cognitive Studies. https://archive.org/details/HowDesignersWork-MakingSenseOfAuthenticCognitiveActivity. Accessed October 05, 2017.

Goffin, K., & Mitchell, R. (2005). *Innovation management: Strategy and implementation using the pentathlon framework*. Houndmills, Basingstoke: Palgrave Macmillan.

Gregory, S. A. (1966a). Design and the design method. In S. A. Gregory (Ed.), *The design method* (pp. 3–31). London: Butterworths.

Gregory, S. A. (1966b). *The design method*. London: Butterworths.

Gregory, S. A. (1966c). *The design method*. Springer US. https://doi.org/10.1007/978-1-4899-6331-4.

Katz, R. (2004). Introduction. In R. Katz (Ed.), *The human side of managing technological innovation: A collection of readings* (2nd ed.). Oxford, UK: Oxford University Press.

King, A. (1993). From sage on the stage to guide on the side. *College Teaching, 41*(1), 30–35.

Laurillard, D. (2012a). *Teaching as a design science: Building pedagogical patterns for learning and technology*. New York/London: Routledge.

Laurillard, D. (2012b). *Teaching as a design science in learning and technology*. Available through https://www.slideshare.net/CdeLondon/cde-ride-dl?next_slideshow=2. Accessed April 2019.

Paper presented at the RIDE 2012 Research and Innovation in Distance Education and eLearning University of London October 2012.
Lazonick, W. (2005). The innovative firm. In J. Fagerberg, D. Mowery, & R. R. Nelson (Eds.), *The Oxford handbook of innovation*. Oxford, UK: Oxford University Press.
Leifer, L. J., & Meinel, C. (2016). Manifesto: Design thinking becomes foundational. In H. Plattner, C. Meinel, & L. Leifer (Eds.), *Design thinking research making design thinking foundational*. Switzerland: Springer.
Lester, R. K., & Piore, M. J. (2004). *Innovation-the missing dimension*. Cambridge, MA: Harvard University Press.
ME310Porto. (2019). *Porto Design Factory*. https://me310porto.com/core-pedagogy/. Retrieved April 2019.
Michaelsen, L. K. (2004). Getting started with team-based learning. In L. K. Michaelsen, A. B. Knight, & L. D. Fink (Eds.), *Team-based learning: A transformative use of small groups in college teaching* (pp. 27–50). Sterling, VA: Stylus Publishing.
Michaelsen, L. K., Knight, A. B., & Fink, L. D. (Eds.). (2004). *Team-based learning: A transformative use of small groups in college teaching*. Sterling, VA: Stylus Publishing.
O'Neill, G. (2015). *Curriculum design in higher education: Theory to practice*. Dublin, Ireland: UCD Teaching & Learning. Available from http://researchrepository.ucd.ie/handle/10197/7137. Accessed December 2019.
OECD. (2005). *Oslo manual: Guidelines for collecting and interpreting innovation data* (3rd ed.). Paris: Organisation for Economic Co-Operation and Development, Directorate for Science, Technology and Industry.
Porto, P. (2019). *Instituto Politécnico do Porto*. https://www.ipp.pt/. Retrieved April 2019.
Pedler, M., Burgoyne, J., & Boydell, T. (1991). *The learning company – A strategy for sustainable development*. Maidenhead, UK: McGraw-Hill.
Poole, M. S., & Van de Ven, A. H. (2000). Towards a general theory of innovation processes. In A. H. Van de Ven, H. L. Angle, & M. S. Poole (Eds.), *Research on the management of innovation: The Minnesota studies* (Vol. 32, pp. 637–662). Oxford, UK/New York: Oxford University Press.
Rothwell, R. (1994). Towards the fifth-generation innovation process. *International Marketing Review, 11*(1), 7–31.
Schumpeter, J. A. (1934). *The theory of economic development: An inquiry into profits, capital, credit, interest, and the business cycle*. Cambridge, MA: Harvard University Press. (1959 printing).
Simon, H. (1996). *The Sciences of the Artificial*. The MIT Press; Cambridge, Massachusetts, third edition.
SUGAR. (2019). *Sugar Design Thinking Global HEI Network managed by the University of St. Gallen*. Avialable at https://sugar-network.org/. Accessed December 2019.
Swedberg, R. (1991). *Joseph A. Schumpeter: His life and work*. Cambridge, UK: Polity Press.
Tushman, M. L., & O'Reilly, C. (2004). The ambidextrous organization: Managing evolutionary and revolutionary change. In M. L. Tushman & P. Anderson (Eds.), *Managing strategic innovation and change: A collection of readings* (2nd ed.). Oxford, UK: Oxford University Press.
Ulrich, K. T., Eppinger, S. D., & Yang, M. C. (2020). *Product design and development* (7th ed.). New York: McGraw-Hill Education.
Van de Ven, A. H. (1986). Central problems in the management of innovation. *Management Science, 32*(2), 590–607.
Verganti, R. (2009). *Design-driven innovation: Changing the rules of competition by radically innovating what things mean*. Boston: Harvard Business Press.
Verganti, R., & Dell'Era, C. (2014). Design-driven innovation: Meaning as a source of innovation. In M. Dodgson, D. M. Gann, & N. Phillips (Eds.), *The Oxford handbook of innovation management*. New York: Oxford University Press.
Wilson, J. Q. (1966). Innovation in organizations: Notes toward a theory. In J. D. Thompson (Ed.), *Approaches to organizational design* (pp. 193–218). Pittsburgh, PA: University of Pittsburgh Press.
Zaltman, G., Duncan, R., & Holbek, J. (1973). *Innovations and organizations*. New York: Wiley.

Part I
Principles

Chapter 2
The Principle of Phronesis

2.1 Introduction

Dante in the Divine Comedy describes his encounter with Aristotle as follows: "I saw the master sage of those who know" Inferno Canto IV, 131. (Dante, 1984 translation)

The role and importance of philosophy continue to be a matter of lively debate within the social science disciplines. Furthermore, according to O'Neill (2015), the educational philosophy of an academic programme is an important statement of the underlying approach to teaching and learning. This section will engage with these debates by proposing a novel philosophical underpinning of design and innovation education in the works of Aristotle. Dante considered Aristotle as epitomising the pinnacle of human reason, and his references to Aristotle came second only to that of the Bible. Without doubt, Aristotle is one of the most influential thinkers in the history of Western civilisation. His works have covered a vast range of subjects such as logic ethics, metaphysics, politics, natural science and physics (ODE, 2006). Aristotle's science was not superseded until the scientific revolution of the sixteenth and seventeenth centuries (ODS, 1999). Critchley (2001, p.7) points out that among other things, ancient philosophy endeavoured to amalgamate knowledge and wisdom: "namely, that knowledge of how things were the way they were, would lead to wisdom in the conduct of one's life." The assumption that ties knowledge and wisdom together is the idea that the cosmos as such expresses a human purpose. This is called the "teleological view of the universe" (Critchley, 2001, p. 7). This section will focus on Aristotle's ethical writings, especially examining his synthesis of the ideas of knowledge and wisdom and their application in practice, which he achieved through the concept of phronesis (*phronēsis*). Taylor (2005) explains phronesis as having, in ancient Greece, connotations of intelligence and soundness of judgement. In his deliberations on ethics, Aristotle used the term to represent the complete excellence of the practical intellect. In the practical sphere, it was the equivalent of *sophia* in the theoretical sphere.

G. J. Costello, *The Teaching of Design and Innovation*, Contemporary Issues in Technology Education, https://doi.org/10.1007/978-3-030-41380-4_2

This chapter argues that the teaching of design and innovation requires a bottom-up approach to "put some meat"[1] on the prevalent theoretical examination of the subject. While much has been written on phronesis, there is a dearth of empirical work on the how the concept can be developed and implemented in practice (which is ironical given the nature of the idea). This work will contribute to this nascent debate by providing empirical evidence, in Chaps. 6 and 7, of embedding Aristotle's phronesis in the curriculum of future managers, in this case engineering and business students, working in cross-functional teams. However, we first must examine the concept in order to justify such a place in the teaching of design and innovation.

The structure of the chapter will be as follows. First there will be a discussion on the philosophical basis for "engaging with the practical". Following this, Aristotle's concept of phronesis is presented. Then there will be an examination of the contemporary topic of phronesis and the modern interest in the area resulting from the work of scholars such as Bent Flyvbjerg and Elizabeth Anne Kinsella. Finally, a discussion of the implications of phronesis for the teaching of design and innovation will be presented.

2.2 Engaging with the Practical

What we must learn to do we learn by doing. (Aristotle)

Praxis (from the Greek *prasso* – doing or acting) concerns the theory of human activity and shares an etymology with pragmatism in the root meaning of *pragma* (deed, affair). According to Schrag (1999), the concept of praxis can be traced back to Aristotle to denote the "sphere of thought and action that comprises the ethical and political life of man, contrasted with the theoretical designs of logic and epistemology (*theoria*)". While the concept of praxis these days is almost completely identified with Marxism, Heidegger in *Being and Time* sought to recover what he considered the real meaning of the term in Aristotle's conception of human praxis (Moran, 2000, p. 196). Hannah Arendt was deeply influenced by Heidegger's lectures on Aristotle and is considered as rehabilitating the notion of praxis in her account of "action" (Moran, 2000, p. 290). The concept of praxis is also considered to be very influential on the development of critical theory which links the term to the "phenomena of discourse, communication, and social practices" (Schrag, 1999). In Eastern Europe the Praxis school published the Marxist journal *Praxis* during the 1960s and 1970s and ran a summer school that attracted Western scholars. The central concern of this group was to "study and influence the role of free creative activity in changing and shaping ethical, social, political, and economic life along humanistic socialist lines" (DeGeorge, 2005). Karol Wojtyla (1979) was an influential phenomenologist who sought to reclaim praxis from Marxism by proposing his own theory of human action. McNerney (2004, p. 128) contends that Wojtyla's

[1] EN Book VI Chapter 7 (Aristotle 1973).

magnum opus *The Acting Person* "can be understood in itself as a restatement of a philosophy of praxis". The question of how to define human praxis is fundamentally based on how you define the human person, and Wojtyla argued that the human person must be regarded as a self-determining subject who is the "efficient cause of his/her own actions" (ibid., p. 128). This contrasts with what he regarded as the dehumanising dogma of Marxism. Furthermore, human activity operates intersubjectively –within a community of persons. Work according to Wojtyla has not just an objective sense such as the transformation of nature into products but has the subjective sense of contributing to self-fulfilment (ibid., p. 159).

In Wojtyla's schema (1979, p. 11), a person revels himself in action:

> [Action] reveals the person and we look at the person through his action...Action gives us the best insight into the inherent essence of the person and allows us to understand the person most fully. We experience man as a person, and we are convinced of it because he performs actions.

Similarly Goldkul (2004, p. 3) emphasises the importance of actions in the influential Pragmatist tradition.

> This does not mean that a pragmatist is only concerned with actions and disregards other issues. A pragmatist researcher lets actions appear as something significant and fundamental to study. The primary concern, following a pragmatist position, in the empirical world is actions.

The purpose of this section was to demonstrate that engaging with the practical has a long history, and I will now turn my attention to Aristotle, the main protagonist that I am proposing for underpinning the teaching of design and innovation.

2.3 Phronesis and Its Origin in the Nicomachean Ethics

> *A man is aware that light meats are easily digested and beneficial to health but does not know what meats are light. Such a man is not so likely to make you well as one who knows that chicken is good for you.* (Aristotle)

Aristotle has his name associated with two treatises on ethics: the Eudemian and the Nicomachean. The former is regarded to have been edited by his pupil Eudemus while the latter by his son Nicomacheus. A third and little know work by Aristotle is called the *Magna Moralia*. The *Nicomachean Ethics* is dated to the last period of Aristotle's life when, in his 50s or 60s, he was head of the Lyceum. Ross (1954) explains that Aristotle regards the purpose of human life as "consisting of the pursuit of ends" (ibid., p. vi) and describes this end as being *eudaimonia*. This concept however is usually translated unsatisfactorily as "happiness", since Aristotle goes beyond this and contends that *eudaimonia* is "an activity of soul in accordance with virtue (ibid., p. vii)". Aristotle describes his idea of virtue as being an activity.

> Just as men become builders by building, they become just by doing just acts, and temperate by doing temperate acts.... And activities which produce excellence are those in which both excess and defect are avoided. (p. viii).

Table 2.1 A taxonomy of the *Nicomachean Ethics* from Ross (1954)

Bk. 1	Deals with the nature of good for man and divides goodness into intellectual excellence and moral excellence
Bk. 2–4	These deal with moral excellence and discuss particular moral virtues in detail. Virtue according to Aristotle being a state of character concerned with choice
Bk. 5	Discusses one of the two cardinal virtues not dealt with in Bks 2–4, namely, justice
Bk. 6	Discusses the various form of intellectual excellence In Aristotle's scheme, phronesis is classified as one of several "intellectual virtues" or "excellences of mind" (Kinsella 2012)
Bk. 7	Earlier part discusses two conditions which lie between virtue and vice – continence and incontinence. The latter part is a discussion of pleasure
Bk 8–9	These are on the subject of friendship and are not regarded as core to the theme of ethics
Bk 10	Elaborates on the concept of *eudaimonia* which is translated by Ross as Well-being

Thompson (1973) however reminds us that Aristotle was keen to take other views into account when developing his ethical programme and that he had a genuine understanding of human nature. Furthermore, statements like "man is a social animal" now seem to be a cliché, but when first proposed by Aristotle, it was quite revolutionary. According to Ross (1954), the plan of the *Nicomachean Ethics* is not simple and was probably delivered as a series of lectures. He provides the following taxonomy of the work (ibid., p. v) which is summarised in Table 2.1.

Chapter 6 of the *Nicomachean Ethics* discusses a number of characteristics of the mind and in particular the two main intellectual virtues *phronesis* (practical wisdom) and *sophia* (philosophic wisdom) (Ackrill, 1973). Furthermore, the person who has phronesis can be described as a *phronimos* having the following attributes (p. 28).

> The *phronimos*[2] is good at deliberation: he can sum up a situation, weigh up various factors, and work out what to do to promote or achieve his objectives. Often enough, because of his experience and wisdom, he can see straight off the best thing to do, without having to go through a process of deliberation.

Taylor (2005) defines phronesis as "comprising a true conception of the good life and the deliberative excellence necessary to realise that conception in practice via choice". In the Nicomachean Ethics, Aristotle describes three approaches to knowledge: *episteme*, *techne*, and *phronesis*. Flyvbjerg (2001) explains that "whereas *episteme* concerns theoretical know-why and *techne* denotes technical know-how, *phronesis* emphasises practical knowledge and practical ethics" (ibid., p. 56). He summarises the three as follows (ibid., p. 57):

- *Episteme – scientific knowledge*. Universal, invariable, context-independent. Based on general analytical rationality. The original concept is known today from the terms "epistemology" and "epistemic".

[2] While the quotation uses *phronimos* as masculine, it goes without saying that the term equally applies to the feminine *phronima*.

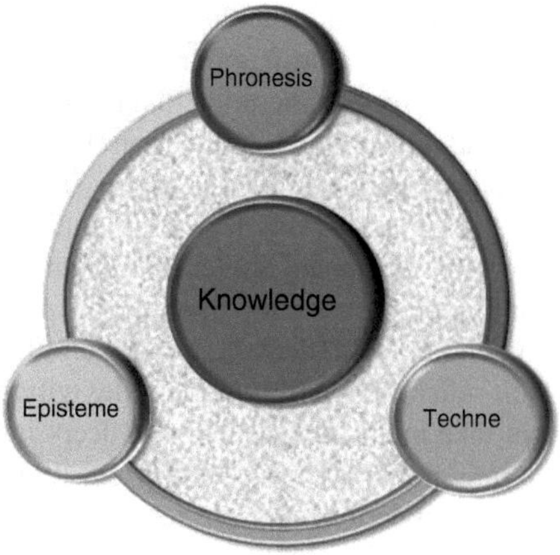

Fig. 2.1 Representation of Aristotle's concept of knowledge

- *Techne – craft art.* Pragmatic, variable, context-dependent. Oriented towards production. Based on practical instrumental rationality governed by a conscious goal. The original concept appears today in terms such as "technique", "technical" and "technology".
- *Phronesis – ethics.* Deliberation about values with reference to praxis. Pragmatic, variable, context-dependent. Oriented towards action. Based on practical value rationality. The original term has no analogous contemporary word (Fig. 2.1).

Aristotle provides the following guidelines for the application of phronesis.

> Observe, too, that prudence (practical wisdom) is something more than knowledge of general principles. It must acquire familiarity with particulars also, for conduct deals with particular circumstances and prudence is a matter of conduct. This accounts for the fact that men who know nothing of the theory of their subject practice it with greater success than others who do know it. It is in fact experience rather than theory that normally gets results. Practical wisdom being concerned with action, we need both kinds of knowledge; nay, we need the knowledge of particular facts more than general principles". *Nicomachean Ethics Book VI Chapter 7*

Flyvbjerg continues to elaborate on the term phronesis in Aristotelian terms:

> In Aristotle's words phronesis is a "true state, reasoned, and capable of action with regard to things that are good or bad for man." Phronesis goes beyond both analytical, scientific knowledge (episteme) and technical knowledge or know-how (techne) and involves judgments and decisions made in the manner of a virtuoso social and political actor. I will argue that phronesis is commonly involved in social practice, and that therefore attempts to reduce social science and theory either to episteme or techne, or to comprehend them in those terms, are misguided. (ibid. p. 14)

Kinsella and Pitman (2012, p. 1) state that their book on phronesis originated from a continuing conversation in which "we voiced concern (bordering on distress) regarding the instrumentalist values that permeate (often without question) our professional schools, professional practices, and policy decisions. Like others, we were grappling with a sense that something of fundamental importance— of moral significance— was missing in the vision of what it means to be a professional, and in the ensuing educational aims in professional schools and continuing professional education." Phronesis, on the other hand, is an intellectual virtue that implies ethics. It involves deliberation that is based on values, concerned with practical judgement and informed by reflection. In thinking about how practitioners might enact phronesis, Kinsella (2012) argues in a related chapter that attention to reflection and judgement is key. Kinsella's study is informed by the seminal reflective practice work of Donald Schön (1983) but suggests a further extension. She proposes a continuum of reflection that informs professional action from (a) receptive or phenomenological reflection, to (b) intentional cognitive reflection, to (c) embodied or tacit reflection, to (d) critical reflexivity. Her analysis acknowledges that reflection can take many forms: it can be deep, interior, emotional and introspective; it can be intentional and based in reason; it may also be tacit, embodied and revealed in intelligent action; and, further, it may be used to critically interrogate assumptions about taken-for-granted understandings in professional life (ibid., p. 4). Phronesis is defined in different ways but usually in ways that imply the significance of reflection, both tacit and explicit, that highlight a relationship to morality and that convey a relationship between reflection and action. Phronesis emphasises reflection (both deliberative and that revealed through action) as a means to inform wise action, to assist one to navigate the variable contexts of practice, and as directed towards the ends of practical wisdom (ibid., p. 35). Because of the importance of reflection in the teaching of design and innovation, I will deal with the subject in some detail in Chap. 5.

Recently, in management scholarship, the influential Andrew Van de Ven (2007) also distinguishes between:

- *Techne* (applied technical knowledge of instrumental or means-ends rationality).
- *Episteme* (basic knowledge in the pursuit of theoretical or analytical questions).
- *Phronesis* (practical knowledge of how to act prudently and correctly in a given immediate and ambiguous social or political situation).

According to Van de Ven, saying that the knowledge of science and practice are different is not to imply that they stand in opposition or they substitute for each other; rather, they complement one another (ibid., p. 3).

2.4 Distinguishing Contemporary from Classical Phronesis

Bent Flyvbjerg describes "phronetic social science" as an approach to the study of social phenomena based on a *contemporary* interpretation of the *classical* Greek concept phronesis, variously translated as practical judgement, practical wisdom, common sense or prudence. Phronetic planning research, an area in which he is influential, is phronetic social science employed in the specific study of policy and planning (Flyvbjerg, 2018).

In the *Sage Dictionary of Qualitative Management Research*, Flyvbjerg points out that phronetic organisational research effectively provides answers to the following four value-rational questions, for specific problems in management and studies:

1. Where are we going with this specific management problem?
2. Who gains and who loses, and by which mechanisms of power?
3. Is this development desirable?
4. What, if anything, should we do about it?

Thus, in his view, phronetic organisational research concerns deliberation, judgement and praxis in relation to the four questions. Praxis is the process by which *phronesis*, as a concept becomes lived reality.

Flyvbjerg, Landman, and Schram (2012) argues that phronesis is commonly involved in social practice and that therefore attempts to reduce social science and theory either to episteme or techne, or to comprehend them in those terms, are misguided. They know that knowledge is power and that they have a civic duty to use this power – and use it wisely, that is, phronetically – in the service of the communities in which they live (S. F. Schram, Flyvbjerg, & Landman, 2013, p. 26). Furthermore, Flyvbjerg (2001) contends that the "principal objective for social science with a phronetic approach is to carry out analyses and interpretations of the status and values and interests in society aimed at social commentary and social action, i.e. praxis". Not without controversy and positioning himself against the influential worldview of Habermas, Flyvbjerg argues that "social and political thinking becomes problematic if it does not contain a well-developed conception of power". Consequently, he focuses on "Michel Foucault's analysis of power as a means of developing a more adequate and contemporary conception of phronesis" (ibid., p. 110).

From the implications of his analysis of power, he adds the following question to the ones above.

5. Who gains and who loses; by which mechanisms of power?

In summary, Flyvbjerg argues that the question of power differentiates contemporary phronesis from classical phronesis. The intense debate in the contemporary literature on the nature of "power", largely influenced by the philosophy of Michel Foucault, is contentious but beyond the scope of this chapter. However, its influence needs to be acknowledged. In a related work, Schram (2012, p. 18) suggests four "interrelated reasons for promoting phronetic social science over the mainstream variety". Flyvbjerg (2001, p. 159) concludes that "reaching the dialogical mode of communication seems crucial to me for practicing phronesis in a democratic society".

2.5 Implications of Phronesis for Educational Practice

Sand (2018, p. 79) describes innovation processes as being "extremely complex and opaque, which makes it tough or even impossible to govern them". Furthermore, he argues against dismissing "the agent-centred approach to responsibility" for the process of innovation. To address this, he has developed a virtue ethical approach. His methodology involves discussing science fiction and biographical literature to elucidate his thesis. This book goes a step further by analysing real-world case studies. The implications of the revival of Aristotle's virtue ethics for teaching design and innovation will be examined in the concluding chapter of the book.

Recently Blok (2018) has requested a new research agenda for the philosophy of innovation stating that this calls for an opening up of the concept of innovation and the philosophical reflection on the topic that is able to address the grand challenges of our time. Furthermore, an objective of this book is to contribute to the debate by providing empirical evidence of embedding Aristotle's phronesis in the curriculum of future managers. To support the goal, I will now discuss concepts from the philosophical literature that emerged from the review above with the objective of developing a taxonomy that will assist the development of phronesis in the education of design and innovation. This classification will be used to examine the case study described in Chap. 6.

2.6 Phronesis and the "Good"

It is thought that every activity, artistic or scientific, in fact every deliberate action or pursuit, has for its object the attainment of some good. Aristotle Nicomachean Ethics Book 1

In Taylor's definition of phronesis provided above, the idea of phronesis is tied to the concept of "a good life". The debate on the importance of the common good is an ancient one, with Cicero's works, *De legibus* and *De republica*, being famous for their assertion of human rights and the brotherhood of man (Gashin, 2005). According to Musa, a major theme throughout the writings of Dante was that the man who does not contribute to the common good fails sadly in his duty (Musa, 1984). Ross (1954, p. vii) proposes, from his work on the Nicomachean Ethics, that "human good turns out to be 'activity of soul' in the sense of being obedient to a rational principle or of apprehending such as principle; in other words, either good moral activity or good intellectual activity".

Jacques Maritain was an important philosopher on the subject of the common good (Maritain, 1973) and was a leading architect of the United Nations Universal Declaration of Human Rights in 1948. Maritain draws heavily on the work of St. Thomas Aquinas who is credited among those who developed Aristotle's thought in the West (Jaffa, 1952). Copleston (1955, p. 59) tells us that "in developing his philosophy Aquinas made considerable use of Aristotle" and "that it is clear that both men developed teleological theories of ethics, the idea of the good being paramount. For both, human acts derive their moral quality from their relation to man's final end" (ibid., p. 198). They also shared the view "that it is the possession of reason which distinguishes man from the animals, to whom he is in many ways similar. It is reason which enables him to act deliberately, in view of a consciously apprehended end, and raises him above the level of purely instinctive behaviour" (ibid., p. 204).

2.7 Phronesis and Hermeneutics

> *Though its sphere is the same as practical wisdom, understanding and practical wisdom are not the same. Practical wisdom issues commands: its end is to tell us what we ought to do and what we ought not to do. Understanding, on the other hand only passes judgment.*
> Aristotle (1978 translation, p. 164)

Malpas (2005) has described Hans-Georg Gadamer as the decisive figure in the development of twentieth-century hermeneutics, and indeed his long life spanned the whole of this period from his birth in 1900 to his death in 2002. An important facet of Gadamer's work was his immersion in the Greek tradition and the programme proposed by Socrates, Plato and Aristotle of pursing philosophy through dialogue and engagement with the *practical.* In fact both Plato and Aristotle saw philosophy as engaging with practice (Moran, 2000, p. 268). Tredennick (1969) points out that Socrates insisted that he was not a teacher but a sort of intellectual midwife who helped "others to bring their thoughts to birth". The Aristotelian concept of phronesis (practical reasoning) is central to the development of hermeneutical understanding in Hans-Georg Gadamer's major philosophical work *Truth and Method* (Dostal, 2002, p. 8). Heidegger's early analysis of Aristotle's phronesis helped Gadamer see that Aristotle's practical philosophy exemplifies integral hermeneutics. According to Lawrence (2002), when Gadamer makes phronesis the heart of his philosophical hermeneutics, he removes all the ambiguity from Heidegger's insight into the relevance of phronesis for a philosophy of human historicity. On the contrary, Gadamer maintains, reading Plato in light of Aristotle's critique, "we see how close the knowledge of the good sought by Socrates is to Aristotle's phronesis" (Zuckert, 2002, p. 212). Whether we are dealing with morality, ethics or politics, Gadamer advocates the necessity for cultivating hermeneutic sensitivity and phronesis in all dimensions of human life. Chapter 1 introduced the topic of interpretation in contemporary literature on design-driven innovation, and the work of Gadamer is relevant to this debate. According to Bernstein (2002), Gadamer prescribes the practical philosophy of Aristotle and his concepts of phronesis as an antidote to the growing pervasiveness of blind technological thinking. The dangers of technological somnambulism for the realm of design and innovation will be examined in Chaps. 3 and 8.

2.8 Phronesis and the Philosophy of Innovation

Recently the philosophy of innovation has been the subject of a special issue of the *Philosophy of Management Journal* edited by Vincent Blok (Vincent Blok, 2018; Hammershøj, 2018; Hühn, 2018; Kamishima, Gremmen, & Akizawa, 2018; Moussavi & Kermanshah, 2018; Sand, 2018; Schlaile, Mueller, Schramm, & Pyka, 2018). The section will provide a brief review of the themes explored in the special issue that are associated with the concept of phronesis. Consequently, these themes will also be used to examine the illustrative case study outlined in Chap. 6.

In his introduction to the special issue, Blok (2018) proposes that the concept of innovation requires philosophical reflection in order to address the grand challenges of our time. Furthermore, the emerging literature on responsible innovation (the main theme of Chap. 3 of this book) adds urgency to this task. I argue that such reflection needs to encompass the area of education and that applying Aristotle's concept of phronesis in the curriculum facilitates such an undertaking. Hühn (2018), in a provocative paper, proposes that Adam Smith's economic view has been largely misinterpreted. Contrary to the canonical "invisible hand" representation of his thought, Hühn, drawing on Smith's *Theory of Moral Sentiments*, proposes that the concept of sympathy (or fellow-feeling) was the main underlying vision of his economic system. Consequently, innovations are useful if their creators constantly imagine that there are placed in the other people's shoes. It should be noted that Smith's *sympathy* is regarded to be nearer the concept of *empathy* as understood in modern parlance, and I will discuss this in more detail in Chap. 3. Hühn contrasts this "Kirkcaldy Smith" from what he calls the twentieth-century fictional "Chicago Smith" and positions Smith as a virtue ethicist with Aristotle being his major influence. Moussavi and Kermanshah (2018) provide a philosophical appraisal on the innovation systems approach which has recently become an important framework in the macro-area of policy-making. Sand (2018) argues that responsible research and innovation presumes rational agents and subsequently develops a virtue ethical approach. Drawing from Aristotle's *Nicomachean Ethics* and using the analogy of teaching an apprentice carpenter, Sand contends that the presentation and discussion of an exemplar is a far better strategy than telling the student exactly what to do. My illustrative case study in Chap. 6 incorporates this type of approach. Kamishima et al. (2018) propose a capability-effectual approach to the very topical innovation of designing robots equipped with artificial intelligence (AI). Importantly for my argument, they contend that such innovation requires the input of stakeholders with the characteristic of phronesis as advocated by Aristotle in the *Nicomachean Ethics*. However, I believe that their contention that AI robots are potentially more intelligent than humans and that robots may evolve into a different species than human is somewhat removed from current reality. Transhumanism raises many fundamental philosophical issues, which, as Hubert Dreyfus eloquently reminds us, are far from being resolved (H. Dreyfus & Dreyfus, 1990). I will take up this discussion in Chap. 8 as I believe it has significant implications for the teaching of design and innovation in the future. Finally, Hammershøj (2018) presents his hypothesis that creativity and innovation are affective processes, which involve transcendence routines that are ultimately uncontrollable and that require leadership rather than management. Such leadership requires enthusiasm to motivate stakeholders such as employees, customers and investors. The initial stages of the innovation, according to Hammershøj, are not founded on cognitive judgement based on logic but affective judgement based on emotions.

Table 2.2 summarises key concepts extracted from the special issue that illuminate the contemporary understanding of Aristotle's notion of phronesis. Furthermore, they will be used to examine the illustrative case study that I will present in Chap. 6.

Table 2.2 Relevant concepts from Philosophy of Management Special Issue (2018)

Author(s)	Concept
Hühn	Virtue ethics
Moussavi and Kermanshah	Learning, interdisciplinary
Sand	Virtue ethics, agent-centred, utilises exemplars
Kamishima, Gremmen and Akizawa	Ethical stakeholder, practical intelligence, effectuation
Hammershøj	Affective processes, transcendence, enthusiasm

Table 2.3 Taxonomy for investigating a phronetic educational artefact (Costello, 2019)

Philosophical concept	Related question
Common good	Does the artefact contribute to the common good and exhibit a sense of responsibility to the wider community?
Power relations	Does the artefact support power as service or power as domination (do some gain and do some lose)?
Hermeneutic	Does the artefact meet the requirement of universal design, i.e. is it understood and accessible to all?
Agency	Does the artefact developer attribute responsibility to persons rather than abstract structures?
Virtue ethics	Does the project result in the development of the character and positive habits of the team members?
Interdisciplinary learning	Does the development of the artefact enable learning through engagement with team members from other disciplines?
Effectuation	Does the project enable team members to attain a feeling of accomplishment?
Affective processes	Does the project develop processes such as emotional intelligence and empathy?
Responsible innovation	Does the innovation demonstrate acceptability, sustainability and societal desirability?

2.9 A Phronetic Taxonomy

The output from a design and innovation process is an artefact of some description: a product or service or something intangible like a business idea or business process. In Table 2.3, I summarise the ideas derived from this chapter into a taxonomy of philosophical concepts and related questions designed to evaluate artefacts that have been developed by student cross-functional project teams working in the area of design and innovation. Furthermore this work addresses the conclusion by Small (2004, p. 194) that at the present time, management programmes include subjects that do not greatly facilitate the development of the ability to think critically.

Small's suggestion for the use of role-play in improvised scenarios to facilitate the learning process has been implemented in the case studies presented in later chapters. For example, in the projects outlined in Chap. 6, it is emphasised that the team must assume the "role" of professionals working on a real-life business scenario.

Here I would like to include an anecdote from the first case study in Chap. 6. It was in the context of a role-play by the students (acting as professionals engaged in a real-life project) presenting their design solutions to three industry professionals. One of the student teams was addressing a very important aspect of design, namely, safety. The students described how they had "fool-proofed" the design in several ways. One of the engineers responded that he never uses the term "fool-proof" in his work in industry. The operators he worked with are very intelligent people and should be respected as such. The term he uses when doing a risk analysis is "fail-safe" rather than "fool-proof". It was a very salient lesson for all the students, and it came from a professional working in a busy manufacturing environment. Afterwards, I thanked the engineer for giving the students such an important and powerful lesson on practical wisdom.

This chapter sets out to explore the following question: *What are the characteristics of a contemporary understanding of phronesis and how can this understanding facilitate its implementation in educational practice?* This undertaking was principally inspired by Aristotle's axiom that "what we learn to do we learn by doing".

2.10 Implications for Theory and the Consequences for Case Studies

Thomas (2011) argues that the case study is concerned with phronesis rather than theory and that this has far-reaching implications for social science research. Here is his summary of phronesis in the context of a case study.

> Phronesis is practical knowledge. It is a model based on personal experience. It is personal, and it helps us make sense of particular situations. (p. 214)

As a result, he contends that there is a pressing need to move away from generalizable knowledge and development of theory associated with the dominant inductive approach. He suggests that instead of talking about theory when examining case studies, we should be talking about phronesis. Drawing on MacIntyre and others, Thomas argues that the prevailing deduction-induction nexus leads to a dead end. He provocatively proposes – after Fish (1990) – that much of what passes for theory in the social sciences is not theory at all but is *theory talk* that is "any form of talk that has acquired cachet and prestige". In its place, he proposed a move towards the "exemplary" knowledge of abduction and phronesis which provides meanings which are "malleable and interpretable in the context of varieties of experience". Is it interesting that Flyvbjerg (2011) makes a similar point in his chapter on the "Case Study" in *The Sage Handbook of Qualitative Research.* Consequently, in Chap. 7, I will propose a schema developed from the case studies of teaching design and innovation in an educational context.

2.11 Conclusions

> *We are what we repeatedly do. Excellence, then, is not an act, but a habit.* (Aristotle)

Engaging with the practical is part of a tradition that goes back to Aristotle "who made frequent reference to concrete examples to illustrate his theoretical points" (Kenny, 2010). Markus and Saunders (2007, p. iv) in their call for more concepts and theories to stimulate management research have specifically requested essays that explore the philosophical foundations on which theory and research is built. The approach of this chapter was to examine the origins of the term phronesis in the ethical writings of Aristotle and to review its influence on contemporary scholarship. The chapter discussed contemporary utilisation of the concepts of "good" and "phronesis" in the relatively recent body of philosophical debate in the management discipline. An emphasis was placed on the work of Bent Flyvbjerg and Elizabeth Anne Kinsella who have separately positioned the concept of "phronetic social science" firmly on the academic agenda. The aim of this chapter was to set out the historical and contemporary understanding of the term phronesis. It provides the groundwork for the implementation of the concept in a teaching context that will be described in Chaps. 6 and 7.

A parting word from Kemmis and Smith:

> A quality of mind and character and action - the quality that consists in being open to experience and being committed to acting with wisdom and prudence for the good. The person who has this virtue has become informed by experience and history and thus has a capacity to think critically about a given situation…and then to think practically about what should be done …When we have phronesis, we are thus prepared…to take moral responsibility for our actions and the consequences that follow from them. The virtue of phronesis is thus a willingness to stand behind our actions. Kemmis and Smith (2008, p. 156) *quoted in* (Kinsella & Pitman 2012, p. 164)

Now I will move on to a chapter on "Responsibility" which, I propose, is a natural progression from the discussion of phronesis grounded in Aristotle's *Nicomachean Ethics*.

Chapter Summary

- Educational philosophy is an important statement of the underlying approach to teaching and learning.
- Engaging with the practical has a long philosophical history dating back to Aristotle.
- In the Nicomachean Ethics, Aristotle describes three approaches to knowledge: *episteme*, *techne* and *phronesis*.
- Phronesis (practical wisdom) is the lesser known of the three approaches but is the most relevant to the teaching of design and innovation.
- There has been a contemporary revival of the concept of phronesis resulting from the work of scholars such as Bent Flyvbjerg and Elizabeth Anne Kinsella.
- A taxonomy was developed from the literature to assist the incorporation of phronesis in the teaching of design and innovation.

References

Ackrill, J. L. (1973). Introduction. In M. Warnock (Ed.), *Aristotle's 'Ethics'/translated from the Greek by W.D. Ross ... and others; selected and with an introduction by J.L. Ackrill*. London: Faber.

Aristotle. (1973). *The ethics of Aristotle, the Nicomachean ethics/translated with an introduction by J.A.K.Thomson* (1953, Trans.). Harmondsworth, Middlesex: Penguin Classics.

Aristotle. (1978, Trans.). *Nicomachean ethics [by] Aristotle. Translated, with introduction and notes, by Martin Ostwald*. Indianapolis, Indiana: Bobbs-Merrill.

Bernstein, R. J. (2002). The constellation of hermeneutics, critical theory and deconstruction. In R. J. Dostal (Ed.), *Cambridge companion to Gadamer*. West Nyack, NY: Cambridge University Press.

Blok, V. (2018). Philosophy of innovation: A research agenda (guest editorial). *Philosophy of Management, 17*(1), 1–5. https://doi.org/10.1007/s40926-017-0080-z.

Copleston, F. C. (1955). *AQUINAS, A Pelican Book*. Penguin Books, Ltd.; 1963 Reprint edition (1955), Harmondsworth, Middlesex.

Costello, G. J. (2019). The philosophy of innovation in management education: A study utilising Aristotle's concept of Phronesis. *Philosophy of Management, 18*(3), 215–230. https://doi.org/10.1007/s40926-018-00104-7.

Critchley, S. (2001). *Continental philosophy : A very short introduction*. Oxford, UK: Oxford University Press.

Dante. (1984, Ttrans.). *The divine comedy: Volume 1: Inferno by Dante Alighieri translated by Mark Musa*. Harmondsworth, Middlesex: Penguin Classics.

DeGeorge, T. T. (2005). praxis. In T. Honderich (Ed.), *The Oxford companion to philosophy* (2nd ed., p. 751). Oxford, UK: Oxford University Press.

Dostal, R. J. (2002). Introduction. In R. J. Dostal (Ed.), *Cambridge companion to Gadamer*. West Nyack, NY: Cambridge University Press.

Dreyfus, H., & Dreyfus, S. (1990). Making a mind versus modelling the brain: Artificial intelligence back at the branch point. In M. A. Boden (Ed.), *The philosophy of artificial intelligence (Oxford readings in philosophy)*. Oxford, United Kingdom: Oxford University Press; 1 edition (July 12, 1990) reprinted 2005.

Fish, S. (1990). *Doing what comes naturally: Change, rhetoric, and the practice of theory in literary & legal studies*. Durham, NC: Duke University Press.

Flyvbjerg, B. (2001). *Making social science matter: Why social inquiry fails and how it can succeed again* (S. Sampson, Trans.). Cambridge, United Kingdom: Cambridge University Press.

Flyvbjerg, B. (2011). Case study. In N. K. Denzin & Y. S. Lincoln (Eds.), *The Sage handbook of qualitative research* (Vol. Chapter 17, 4th ed., pp. 301–316). Thousand Oaks, CA: Sage.

Flyvbjerg, B. (2018). *What is Phronesis and Phronetic social science?* https://www.linkedin.com/pulse/what-phronesis-phronetic-social-science-bent-flyvbjerg-%E5%82%85%E4%BB%A5%E6%96%8C-/. Accessed December 2018.

Flyvbjerg, B., Landman, T., & Schram, S. (Eds.). (2012). *Real social science: Applied phronesis/edited by Bent Flyvbjerg, Todd Landman, Sanford Schram*. Cambridge, United Kingdom: Cambridge University Press.

Gashin, J. C. A. (2005). Cicero, Marcus Tullium. In T. Honderich (Ed.), *The Oxford companion to philosophy* (2nd ed.). Oxford, UK: Oxford University Press.

Goldkuhl, G. (2004). *Meanings of pragmatism: Ways to conduct information systems research*. Paper presented at the 2nd International Conference on Action in Language, Organisations and Information Systems (ALOIS-2004), 17–18 March, 2004, Linköping University, Sweden.

Hammershøj, L. G. (2018). Conceptualizing creativity and innovation as affective processes: Steve Jobs, Lars von Trier, and responsible innovation. [journal article]. *Philosophy of Management, 17*(1), 115–131. https://doi.org/10.1007/s40926-017-0053-2.

Hühn, M. P. (2018). Responsible innovation: A Smithian perspective. [journal article]. *Philosophy of Management, 17*(1), 41–57. https://doi.org/10.1007/s40926-017-0057-y.

Jaffa, H. V. (1952). *Thomism and aristotelianism: A study of the commentary by Thomas Aquinas on the Nicomachean ethics*. Cambridge, UK: University of Chicago.

Kamishima, Y., Gremmen, B., & Akizawa, H. (2018). Can merging a capability approach with effectual processes help us define a permissible action range for AI robotics entrepreneurship? [journal article]. *Philosophy of Management, 17*(1), 97–113. https://doi.org/10.1007/s40926-017-0059-9.

Kemmis, S., & Smith, T. (2008). Personal praxis: Learning through experience. In P. S. P. Salo & S. Kemmis (Eds.), *Enabling praxis: Challenges for education* (2nd ed., pp. 15–36). Rotterdam, The Netherlands: Sense Publishers.

Kenny, A. (2010). *A new history of western philosophy*. Oxford, UK: Oxford University Press.

Kinsella, E. A. (2012). Practitioner reflection and judgement as Phronesis. In E. A. Kinsella & A. Pitman (Eds.), *Phronesis as professional knowledge: Practical wisdom in the professions*. Rotterdam, NLD: Sense Publishers.

Kinsella, E. A., & Pitman, A. (Eds.). (2012). *Phronesis as professional knowledge: Practical wisdom in the professions*. Rotterdam, NLD: Sense Publishers.

Lawrence, F. (2002). Gadamer, the hermeneutic revolution, and theology. In R. J. Dostal (Ed.), *Cambridge companion to Gadamer*. West Nyack, NY: Cambridge University Press.

Malpas, J. (2005). Hans-Georg Gadamer. In E. N. Zalta (Ed.), *The stanford encyclopedia of philosophy* (Winter 2014 ed.). Available through http://plato.stanford.edu/archives/win2014/entries/gadamer/. Accessed December 2019.

Maritain, J. (1973). *Person and the common good*. Notre Dame, Indiana: University of Notre Dame Press.

Markus, M. L., & Saunders, C. (2007). Looking for a few good concepts…and theories…for the information systems field. *MIS Quarterly, 31*(1), iii–vi.

McNerney, J. (2004). *John Paul II: Poet and Philosopher*. Burns & Oates Ltd. London.

Moran, D. (2000). *Introduction to phenomenology*. London/New York: Routledge.

Moussavi, A., & Kermanshah, A. (2018). Innovation systems approach: A philosophical appraisal. [journal article]. *Philosophy of Management, 17*(1), 59–77. https://doi.org/10.1007/s40926-017-0078-6.

Musa, M. (1984). Introduction. In *The divine comedy: Volume 1: Inferno by Dante Alighieri translated by Mark Musa*. Harmondsworth, Middlesex: Penguin Classics.

O'Neill, G. (2015). *Curriculum design in higher education: Theory to practice*. Dublin, Ireland: UCD Teaching & Learning. Available from http://researchrepository.ucd.ie/handle/10197/7137. Accessed December 2019.

ODE. (2006). Aristotle. In *Oxford Dictionary of English* (2nd ed., Revised). Oxford, UK: Oxford University Press.

ODS. (1999). Aristotle. In J. Daintith & D. Gjertsen (Eds.), *Oxford dictionary of scientists* (2nd ed.). Oxford, UK: Oxford University Press.

Ross, D. (1954). Introduction. In *The Nicomachean ethics: translated and introduced by Sir David Ross*. Oxford, United Kingdom: Oxford University Press.

Sand, M. (2018). The virtues and vices of innovators. [journal article]. *Philosophy of Management, 17*(1), 79–95. https://doi.org/10.1007/s40926-017-0055-0.

Schlaile, M. P., Mueller, M., Schramm, M., & Pyka, A. (2018). Evolutionary economics, responsible innovation and demand: Making a case for the role of consumers. [journal article]. *Philosophy of Management, 17*(1), 7–39. https://doi.org/10.1007/s40926-017-0054-1.

Schön, D. A. (1983). *The reflective practitioner: How professionals think in action*. New York: Basic Books.

Schrag, C. O. (1999). Praxis. In R. Audi (Ed.), *The Cambridge dictionary of philosophy* (p. 731). New York: Cambridge University Press.

Schram, S. (2012). Phronetic social science: An idea whose time has come. In B. Flyvbjerg, T. Landman, & S. Schram (Eds.), *Real social science: Applied phronesis*. Cambridge, United Kingdom: Cambridge University Press.

Schram, S. F., Flyvbjerg, B., & Landman, T. (2013). Political science: A phronetic approach. *New Political Science, 35*(3), 359–372. https://doi.org/10.1080/07393148.07392013.07813687, http://bit.ly/07393115A07393140kEL.

Small, M. W. (2004). Philosophy in management: A new trend in management development. *Journal of Management Development, 23*(2), 183–196. https://doi.org/10.1108/02621710410517265.

Taylor, C. C. W. (2005). Phronēsis. In T. Honderich (Ed.), *The Oxford companion to philosophy* (2nd ed.). Oxford, UK: Oxford University Press.

Thomas, G. (2011). *How to do your case study: A guide for students and researchers*. Thousand Oaks, California: SAGE Publications.

Thompson, J. A. K. (1973). Introduction. In *The ethics of Aristotle, the Nicomachean ethics/ translated by J.A.K.Thomson* (1953, Trans.). Harmondsworth, Middlesex: Penguin Classics.

Tredennick, H. (1969). Introduction. In *Plato: The last days of Socrates*. Harmondsworth, Middlesex: Penguin Classics.

Van de Ven, A. H. (Ed.). (2007). *Engaged scholarship: A guide for organizational and social research*. Oxford, UK/New York: Oxford University Press.

Wojtyla, K. (1979). *The acting person (issue 10 of Analecta Husserliana)*. Dordrecht, Netherlands: D. Reidel Publishing Company.

Zuckert, C. H. (2002). Hermeneutics in practice: Gadamer on ancient philosophy. In R. J. Dostal (Ed.), *Cambridge companion to Gadamer*. West Nyack, NY: Cambridge University Press.

Chapter 3
The Principle of Responsibility

3.1 Introduction

The relationship of business ethics to design and innovation has been recently explored in empirical work carried out in Finland (Pučėtaitė, Novelskaitė, Lämsä, & Riivari, 2016; Riivari & Lämsä, 2014; Riivari, Lämsä, Kujala, & Heiskanen, 2012) with the conclusion that ethical organisational culture affects organisational innovativeness. Blok (2014, p. 1) maintains that the grand challenges of these times warrant "extensive reflection on the impact of innovations" and the need to deliberate on the "different purposes of, and motivation behind, innovations". This task is the primary focus of the nascent research area of responsible innovation (RI). Furthermore, *responsibility* is an important concept in the study of moral and ethical philosophy and hence the title of this chapter (Klein, 2005). The main lens through which this chapter looks at design and innovation is that of phenomenology with particular reference to Edith Stein's work on empathy and Emmanuel Levinas' work on the concept of the "Other" in order to facilitate dialogue between philosophers and academics teaching design and innovation. The chapter examines the stakeholder engagement for responsible innovation (SEiRI) theoretical framework proposed by Blok, Hoffmans and Wubben in light of the notions of *empathy* and *Other*. The objective is to understand and explain what is meant by von Schomberg's definition of RI as a process of *mutual responsivity* of actors. Furthermore, it proposes the work of philosopher Edith Stein as a salient contribution to the debate on responsible research and innovation (RRI). Note that the emerging literature uses the terms responsible innovation (RI) and responsible research and innovation (RRI), and I will use both terms interchangeably. I am conscious of Stein's comment that "philosophy is walking on the edge of the abyss", but I believe that undertaking this debate can prevent the trajectory of future design and innovation output from going over the aforesaid abyss. The chapter is organised as follows. First, in order to provide a primer on the subject, a review of literature on responsible innovation is presented. The subject of empathy is then explored. Following this, an introduction

G. J. Costello, *The Teaching of Design and Innovation*, Contemporary Issues in Technology Education, https://doi.org/10.1007/978-3-030-41380-4_3

is given to a major philosopher in the phenomenological movement, Edith Stein, who has much to offer the current debate on the responsible use of technological innovation. Then the implications of empathy for the SEiRI theoretical framework are explored. Finally, I will present the phenomenology of Emmanuel Levinas as an accompanying lens to influence the development of the literature on responsible research and innovation.

3.2 Responsible Innovation

According to Stilgoe (2013), the "broad aim of responsible innovation is to connect the practice of research and innovation in the present to the future that it promises and helps to bring about". Furthermore Owen, Heintz, and Bessant (2013) argue that responsible innovation challenges us to ask what kind of future we want from science and innovation and the values this is based on. Additionally, they insist that RI is not just a question of being another ethical review or bureaucratic hurdle but a positive realignment of innovation to socially desirable ends, which is values-based rather than just rules-based. von Schomberg (2013) states that responsible research and innovation (RRI) is becoming an increasingly important factor in European policy, in particular in the most recent EU research framework "Horizon 2020" which sets out the strategy for the EU to become an Innovation Union. He makes the case that RRI "should be understood as a strategy of stakeholders to become mutual responsive to each other and anticipate research and innovation outcomes underpinning the 'grand challenges' of our time for which they share responsibility (p. 1)." He proposes the following definition for RRI (ibid., p. 19).

> *Definition*: Responsible Research and Innovation is a transparent, interactive process by which societal actors and innovators become mutually responsive to each other with a view to the (ethical) acceptability, sustainability and societal desirability of the innovation process and its marketable products (in order to allow a proper embedding of scientific and technological advances in our society).

One of the main challenges of von Schomberg's definition is to elucidate what is meant by "societal actors and innovators *become mutually responsive* to each other". A central argument of this chapter is that the concept of empathy can inform the quest to understand and explain what is meant by *mutual responsivity* of actors. Furthermore, it contributes to the development by Blok (2014, p. 4) of the concept of "stakeholder dialogue" in RRI and on the question of "the communication process" involved. In addition, Blok and Lemmens (2015) call for a radical transformation of the very concept of innovation in light of the emerging RI debate. They provide their definition of RI in the arrangement of a formula:

> Responsible innovation = regular innovation + stakeholder involvement with regard to ethical and societal aspects

Blok and Lemmens state that in order to form these ethical and societal aspects, stakeholders should be "*empathetic* with regard to the interests of others, to take the perspective or standpoint of the *other*" (ibid., p. 31). Exploring these concepts is an important motivation for this chapter. They also comment that "corporate innovation is underrepresented in current research on the concept of responsible innovation" (p. 30). Furthermore, Blok, Hoffmans, and Wubben (2015) in their examination of the critical issues and management practices associated with stakeholder engagement for responsible innovation propose the SEiRI theoretical framework consisting of four characteristics:

- *Transparency* – Coping with threats to competitive advantage, lack of control and fear of knowledge leakage.
- *Interaction* – To overcome critical issues, companies and stakeholders could interact by engaging in stakeholder dialogue. Different visions, goals, motives, sectors, values and power imbalances need to be managed.
- *Responsiveness* – Same concerns as with "interaction" above.
- *Co-responsibility* – Considering the investment decision being the responsibility of the investor alone. In addition, co-responsibility may result in legitimacy losses by NGOs.

Blok et al. conclude that "the results of this research are a first contribution to the finding new ways to engage stakeholders in RI" (ibid., p. 162). These attributes will be explored in this chapter through the lens of Edith Stein's phenomenology of *empathy* and Emmanuel Levinas' concept of the *Other* in order to respond to this challenge.

3.3 Empathy as an Enabler of Responsibility

According to the Oxford Dictionary of English (ODE, 2006b) empathy is defined as "the ability to understand and share the feelings of another". Furthermore, it points out that there can be a great deal of confusion between the words *empathy* and *sympathy* (p. 569). Sympathy on the other hand means "feelings of pity and sorrow for someone else's misfortune". Richmond (2005) states that empathy is sometimes prompted by imaginative exercises such as "stepping into someone's shoes". Gordon (1999) proposes that many of the concepts of empathy were in fact discussed under the general term "sympathy" by Hume, Adam Smith and other Scottish philosophers. The word was introduced to the English language in the early twentieth century as a rendering of the German word *Einfühlung* (feeling into) in the literature on aesthetics. The concept was given its classical formulation in the work of Theodor Lipps who characterised it "as a process of involuntary inner imitation whereby a subject identifies through feeling with the movement of another body" whether that be real or illusory (Makkreel, 1999, p. 255). Stein (1989, p. 12) explains that "Lipps depicts empathy as an 'inner participation' in foreign experiences". Currently the concept is utilised in debates on moral psychology, the imagination and the simulation/theory debate. *Imagination* is a major theme in the design and innovation literature and thus calls for further exploration of the concept of empathy in the field.

Furthermore, Gordon (1999, p. 261) concludes that empathy "has been considered a pre-condition of ethical thinking and a major contributor to social bonding and altruism, mental state attribution, language use and translation". The concept of empathy has been divided by Darwall (1998) into three categories:

(i) *Emotional Contagion*. The most rudimentary form of empathy is "emotional contagion" or "infection" as when one "catches" a feeling or emotional state from another, not by imaginative projection, but more directly. Walking into a room filled with laughter and convivial conversation, we feel differently than when the room is filled with depression (or with tension). This is a form of what Hume called "sympathy", which he defined as the "propensity we have… to receive by communication [the] inclinations and sentiments" of others, and when he said, quoting Horace, that "the human countenance…. borrows smiles or tears from the human countenance".
(ii) *Projective Empathy and Simulation*. Projective empathy is not simply copying others' feelings or thought processes as we imagine them. Rather, we place ourselves in the other's situation and work out what to feel, as though we were they. This puts us into a position to second the other's feeling or dissent from it. As [Adam] Smith puts it, we thereby express our sense of the "propriety" of the other's feeling, whether, that is, we think it warranted or not.
(iii) *Proto-Sympathetic Empathy*. Like genuine sympathy, proto-sympathetic empathy has the other person and his plight as object. When we imagine what another person's grief is like for him, we are focused on the other person and his grief. And this means that the distress we feel vicariously by projective identification can find a new target, namely, his distress, thereby giving rise to sympathy.

3.4 Empathy and Technology

Whatever difference of opinion exists on the nature and our relationship with innovation and technology, there is almost universal agreement that it is a very important topic. In the opening years of the third millennium, innovation and technology is ubiquitous with its influence becoming even more prevalent with developments such as artificial intelligence (AI). I will consider the implications further in the concluding chapter. Kline (2002) "suggests that we think of technology first and foremost as involving both sociotechnical systems of manufacture and sociotechnical systems of use". The role of technological innovations in society has gathered considerable momentum in the last 60 years in ways that, for example, Heidegger (1977) would not have dreamed. Heise (2004) sees the progress in science and technology as "contributing to a sense of a new historical age" (p. 136).

According to the economist John Kenneth Galbraith (1967), technology means the systematic application of scientific or other organised knowledge to practical tasks. The Oxford Dictionary of English defines technology as the application of scientific knowledge for practical purposes (ODE, 2006a). Technology is used "to represent things, action, processes, methods and systems" (Kline, 2002, p. 210). Innovation can have a wide range of meanings from innovations in physical objects to software to creative arts and to literature. There is widespread agreement that technology is now an integral part of human existence and crucial to the development and operation of a modern society. However, my review of innovation definitions developed over the last 50 years (Costello, 2015) indicates a lack of engagement with ethics in the mainstream innovation literature. This is despite noteworthy scholarship on the philosophy of technology and responsible innovation carried out in institutes such as the Department of Values, Technology & Innovation (VTI) at Delft University of Technology (VTI, 2019) by researchers such as Martin Sand (2018). He argues that responsible research and innovation presumes rational agents and subsequently develops a virtue ethical approach. Drawing from Aristotle's *Nicomachean Ethics* and using the analogy of teaching an apprentice carpenter, Sand contends that the presentation and discussion of an exemplar is a far better strategy than telling the student exactly what to do.

In conventional innovation literature, any technological innovation is often assumed to be good, which is very problematic, when you consider the case of an "innovative" landmine. Technology influences every aspect of contemporary human life. It affects social and ecological systems, and it is a "source of power, vulnerability, and inequality"(Sandler, 2013). How we relate to others as persons in a technological society is an important subject, so I will address this by looking for some answers from the world of phenomenology. However, in this age of avatars and the Internet of things (IoT), I believe that there is much work to be done on examining the implications for empathy from virtual worlds and social networks. In her research at Harvard, Dorothy Leonard proposes emphatic design as the best method to import knowledge from the market place (D. A. Leonard & Rayport, 1997). There has been a significant move from product to service innovation over the last number of years, and Pitt, Watson and Kavan (1995, p. 177) claim that empathy is one of the "five dimensions that are used by customers when evaluating service quality, regardless of the type of service". Also the importance of empathy in customer care and the dangers of technology impeding such a relationship has been stressed by Gorry and Westbrook (2011). There is growing interest in the area of virtual machine interaction (Potapova & Rodionov, 2014) and innovations such as embodied conversational agents (ECAs) where "virtual agents expressing empathic emotions toward users have the potentiality to enhance human-machine interaction" (Niewiadomski, Ochs, & Pelachaud, 2008, p. 37). This however is a controversial area requiring debate on the ethical issue surrounding avatars. The last point is of particular importance in research that involves vulnerable populations such as children and people with special needs (Pettersson, 2002). In the area of human resource practices, empathy was found to be important in the "communication of clear expectations and goal alignment throughout the organization" (Florea, Cheung, & Herndon, 2013, p. 401). In a macro-economic study, Czap et al. (2012, p. 642) propose that there is an important "role for empathy-sympathy (the basis for an internalised, shared other interest) in tempering and conditioning the more fundamental tendency to pursue self-interest". This is important to the debate on the structures and interrelationships of the economic and social system, in the wake of spectacular eruptions within both Marxism and neoliberalism in the recent past. Staying at the macro, even lofty level, Zakaras (2003, p. 497) investigates Isaiah Berlin's "conception of empathy and its close connection to his hopes for human freedom". Later, I will investigate the applicability of the concept of empathy to the stakeholder engagement for responsible innovation (SEIRI) as proposed by Blok et al. (2015). Before this however, I will introduce Edith Stein, a major influence on the development of phenomenology and especially the subject of empathy.

3.5 Edith Stein: Philosopher of Empathy

Edith Stein's doctoral thesis "On the Problem of Empathy" was completed under Edmund Husserl in the University of Freiburg in 1916 and awarded "summa cum laude" (W. Stein, 1989). Stein was Husserl's protégé but being a woman of Jewish origin was unable to obtain a university position because of the ideological intolerance of that time. Her doctoral thesis was written during the atheistic phase of her life, but it is interesting that she analysed empathy in the context of the complete psycho-physical-spiritual person (E. Stein, 1989). Here is he own account of how her research on empathy (German *Einfühlung*) resulted from a lecture given by the "master himself" (Teresia de Spiritu Sancto, 1952).

> Husserl in his course on Nature and Spirit had maintained that an objective external world can only be experienced inter-subjectively (i.e. by a plurality of individual knowing subjects) who are in a position to exchange information with each other; which means that such an experience presupposes other individuals. Husserl, following Theodor Lipps, named this experience "empathy", but did not explain what it consisted of. Here was a gap which was worthwhile filling; I wanted to discover what empathy meant.

It should be noted that some commentators point out that the German word *Geist*, as used by these philosophers, is not accurately translated as "Spirit" which has a mainly religious semantic in the English language. W. Stein states that the German understanding of *Geist* is somewhere between the terms mind and soul and its philosophical study deals with the creative human spirit. For example, Scheler included such concepts as beauty in his examination of the spiritual. Edith Stein's later life was dramatic both as feminist and as a Carmelite where she continued to correspond with leading phenomenologists and to publish in the Journal of Phenomenology until her death in the gas chamber of Auschwitz in 1942. Martin Heidegger invited Stein to contribute to a special edition of the *Jahrbuch für Philosophie und phänomenologische Forschung* to mark Husserl's seventieth birthday for which she contributed her famous paper "*An attempt to contrast Husserl's Phenomenology and the Philosophy of St. Thomas Aquinas*" (Stein, 1993). Max Scheler was another important influence on Stein, and she was indebted to his insistence on "bracketing", the exercise of which challenged her to suspend every form of a priori prejudice and contributed to her *empathising* with other cultures and beliefs (Posselt, 2005). Both philosophers were somewhat disappointed by the Husserl's tendency towards idealism in his later work and continued to identify themselves with the realism of the early Husserl. In Stein's political thought, "any state exists only for the benefit of human beings", and she was convinced that "humanity is fundamentally one community, precious beyond measure (ICS, 2007). Stein's academic relationship with Heidegger and his philosophy is problematic, and this is treated in some detail by Nota (1988). According to Ameriks (1977, p. 104), "Stein produced a revised and coherent draft of Husserl's manuscripts on time-constitution, a draft that Heidegger later published in his own name as editor and no mention of Stein". This is presented in note 14 on page 8 of *Philosophy of Psychology and the Humanities* (Stein, 2000).

Edith Stein was an influential feminist in Germany between the two World Wars. Her writings on the role of women are significant and worthy of further investigation (Stein, 1996). She proposes that empathy is especially a gift of women and enables "her to participate, understand and stimulate" in her role as worker, interpreter and teacher (p. 115).

Now I will examine Edith Stein's writings on empathy in more detail given its importance for this chapter.

3.6 Stein on Empathy

Moran (2000) concludes that Stein's doctoral dissertation on empathy represents a dependable guide to Husserl's thinking on the subject which had been alluded to in his work *Ideas II.* Empathy for both philosophers was a "non-primordial experience which reveals a primordial experience. Empathy is not a matter of judgement, reasoning or ideation in general. It is a *founded* experience (ibid., p. 176)". Furthermore, the concept was integral to Husserl's examination of intersubjectivity and the experience of the other (*Frenderfahrung*). According to Moran (2000, p. 176), he radicalized the problem in the *Fifth Cartesian Mediation.* "The problem is not: how do I understand the other? Rather Husserl's problem is: how is the other *constituted* for me?" Basehart (1989, p. x) speaks of Edith Stein's conviction that "phenomenology was the most appropriate approach to the investigation of the structure of the human person and she gave it her best efforts during her scholarly career". But the act of empathising with the other has also important consequences for understanding our own nature as well as that of others, as we become "aware of the levels of value in ourselves by empathizing with persons of our own type" (W. Stein, 1989, p. xxiii). The idea of empathy being a two-way process is captured in Fig. 3.1.

In her doctoral thesis *On the Problem of Empathy*, Stein explained her way of thinking: "The world in which we live is not only a world of physical bodies but also of experiencing subjects external to us, of whose experiences we know". Furthermore the "individual is not given as a physical body, but as a sensitive, living body belonging to an 'I', an 'I' that senses, thinks, feels and wills" (E. Stein, 1989, p. 5). However we should remain cognisant of Sawicki's observation that the English word *empathy* has a narrower and somewhat different connotation than the German *Einfühlung*

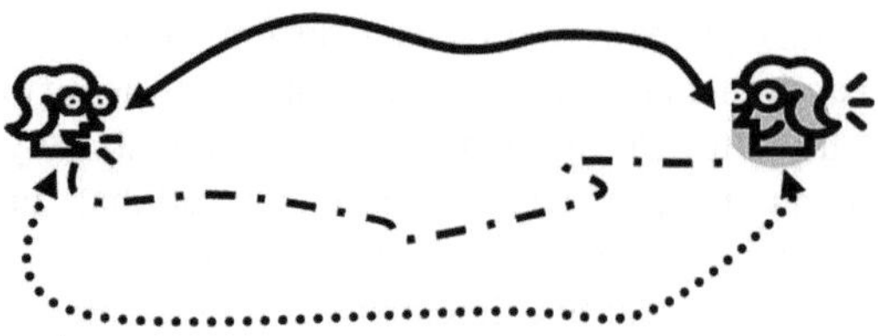

Fig. 3.1 Empathy as a two-way process

– see note 128 on page 93 of Stein (2000). In the next section, I will discuss the implications of Stein's work on empathy for responsible innovation using the SEiRI framework, proposed by Blok, Hoffmans and Wubben.

3.7 A Real-World Example of the Importance of Empathy

I will now include an anecdote for my research in the area of design and innovation which aims to provide a concrete example of empathy in the world. It is taken from the industry case study that is presented in Chap. 6 (Costello, Cresham, & Donnellan, 2007). The objective is to show that the topic has relevance in the very pragmatic world of business. The quotes are taken from an interview with a member of Manufacturing Engineering Department in APC, Ireland. The engineers and technicians in the department define, design and implement the way in which the product is built. Responsibilities include writing the work instructions, installation of lines, fixtures and machines, planning capacity and headcount, ergonomics and health and safety (H&S). A Lean project, undertaken in the plant, was seen to introduce more flexibility within the organisation (Womack & Jones, 2003). The cross-functional teamwork and cooperative decision-making were especially beneficial to manufacturing engineering who traditionally made their own decisions and then presented them to the production department as a fait accompli. As a result, in many occasions the projects exposed their work to retrospective criticism. During the interview, the subject of empathy emerged unexpectedly and without prompting from the interviewer. This is what one of the technicians said:

> Empathy is seeing someone in the situation they are in, especially in cases where they are striving to achieve a goal and are hindered by circumstance over which they have no control. Then it is important to have colleagues that are supportive and available to sort out a problem. Very often it is the conscientious person that will try and try again to make things right without raising the issue. The root cause may be a bad process and this can result in de-motivation of the person involved. The culture must change from a situation where a person has to say "*I* have a problem" to "*We* have a problem" and where the response of the team is "How can *we* help you".
>
> The reality of working in a fast changing supply chain environment demands that people have the human quality and attitude of being able to cope with imperfect or messy processes and deliver for the team. However, the "strive for perfection" proposed by Lean thinking means that such processes which hitherto might just be accepted are now highlighted as an opportunity for improvement (kaizen).

On the subject of decision-making in an "empathetic" environment, he continued:

> Decisions made within this environment should be made with the attitude of imagining that you were making the decision in your own company. If I had to choose members of a project team, one of the main characteristics that I would look for when picking a person would be "empathy".

I hope that this excerpt from a real-world case study indicates that the concept of empathy has significance in the lives of people in a busy working environment. I will now return to a more theoretical consideration of the subject, albeit with the aim to provide guidance in real-life situations.

3.8 Empathy and the SEiRI Theoretical Framework

The four characteristics of SEiRI, outlined earlier, will be viewed through the phenomenological lens of Edith Stein as presented in her doctoral thesis (Stein, 1989). The argument is that a philosophical underpinning is required for the theoretical framework of Blok et al. and Stein's work on empathy provides such a basis. Furthermore, it will provide impetus for an examination of the role and implementation of empathy in the phases of a design and innovation process: the first phase consisting of discovery and scoping and the middle phase consisting of the development and associated business case, together with the final phase of testing/validation and launch (Blok et al., 2015, p. 156). Stein's work facilitates the understanding and hence removal of the barriers associated with much stakeholder interaction. This consideration is especially important given the radical redefinition of design and innovation in an increasingly "open" world (Chesbrough, 2003).

3.9 Transparency

Stein (1989, p. 33) proposes that empathy "proves to have yet another side as an aid to comprehending ourselves" since "people are generally inclined to ascribe to themselves better motives than they actually have and are not conscious of many of their emotional impulses at all". This pertains to the problem identified by Scheler that inner perception "contains within it the possibility of deception".

> Empathy now offers itself to us as a corrective for such deceptions along with further corroboratory or contradictory perceptual acts. It is possible for another to "judge me more accurately" than I judge myself and give me more clarity about myself. For example, he notices that I look around me for approval as I show kindness, while I myself think I am acting out of pure generosity. This is how empathy and inner perception work hand in hand to give me myself to myself. (ibid., p. 89)

Consequently, empathy is important in understanding our own nature as well as others, particularly in the context of teamwork, which is integral to the design and innovation process.

3.9.1 Interaction

Edith Stein analysed empathy in the context of the complete psycho-physical-spiritual person. As her work developed, it became more and more focused on the human person, not as an isolated ontological individual, but relating to other people in a community. In her discussion of the essence of the acts of empathy, she treats empathy in terms of the constitution of the psycho-physical individual. She also argued that there is an important link between empathy and knowledge.

> If we take the self as standard, we lock ourselves into the prison of our individuality. Others become riddles for us, or still worse, we remodel them into our image and so falsify historical truth. (ibid., p. 116)

The development of knowledge and introducing it into practice is central to the design and innovation process. Empathy not only contributes to responsibility for the design and innovation, but it is fundamental to the fruitfulness of the innovation process itself.

3.9.2 Responsiveness

Our responsiveness to others and the world in which we exist and work is very much dependent on empathy. The same world is not merely presented now in one way and then in another (ibid., p. 64).

> Thus, empathy as the basis of intersubjective experience becomes the condition of possible knowledge of the existing outer world, as Husserl, and also Royce present it. (ibid., p. 64)

The way we respond to others is important on how we conceptualise value.

> Every comprehension of different persons can become the basis of an understanding of value. Since, in the act of preference or disregard, values often come to givenness that remain unnoticed in themselves, we learn to assess ourselves correctly now and then. We learn to see that we experience ourselves as having more or less value in comparison with other. (ibid., p. 116)

Such responsiveness is integral to the maturity of design and innovation as a responsible process whose success is mutually dependent on the ethical responses inherent in the development of the design and the implementation of the innovation.

3.9.3 Co-responsibility

Stein states that we become "aware of levels of value in ourselves by empathising with persons of our own type" (Stein, 1989, p. xxii). Furthermore, she contends that "mechanical causation as an explanation of physical phenomena is not appropriate for explaining *spiritual* phenomena" (Stein, 1989, p. xxiii) (see above comment on *Geist*).

> The world in which we live is not only a world of physical bodies but also of experiencing subjects external to us of whose experiences we know.

However, she distances herself from psychology's tendency to extreme subjectivism where:

> the explained phenomenon becomes a "subjective creation" without "objective meaning". We cannot accept this interpretation.

Furthermore, according to Sawicki (2000, p. xi), Edith Stein's twin treatises penned shortly after the end of the First World War entitled *Sentient Causality* and *Individual and Community* propose a theory of human creative achievement which is relevant to the discipline of responsible innovation given the importance of creativity to its process. "The 'sentient causality' which is operative within each human being is connected into the 'individual and community' power reservoirs to which all persons contribute and upon which all persons draw. In other words, she connects psyche and mind (ibid., p. xiii)". These concepts especially in the context of community and the co-responsibility of stakeholders to responsible innovation can provide fertile ground for the progress and teaching of RI.

Now I will take a related theme from the influential ethical work of Emmanuel Levinas to complement and add to Edith Stein's formulation of empathy.

3.10 Emanuel Levinas: Philosopher of the "Other"

Davis (1996, p. 1) has described the philosophy of Emmanuel Levinas as being permeated by one simple but profound theme: "Western Philosophy has consistently practiced a suppression of the *Other*". The principal objective of this section is to complement Stein's work on empathy by examining the notion of the "face" of design and innovation through the lens of Levinas. Born in Lithuania, Emmanuel Levinas is credited with introducing phenomenology to France after his studies with Husserl and Heidegger, who were major influences on his work. His considerable volume of publications over a period of 60 years made him a key catalyst in the development of French twentieth-century philosophy including the emergence of the existentialism of Jean-Paul Sartre and Maurice Merleau-Ponty (C. Davis, 1996, p. 3). His major work *Totality and Infinity (1961)* explored such themes as time, death and relations with others, and increasingly his philosophy was concerned with marking out an ethical "face-to-face" with the Other: a concept that while "immediate and singular, is none the less transcendent" (Ainley, 2005, p. 512). Levinas gave phenomenology a radical ethical orientation variously described as his "phenomenology of alterity" (*alterité*) or his "phenomenology of sociality" that starts from the experience of "the face" (*le visage*) and emanates from the other's "proximity". The philosopher Alphonso Lingis (1998, p. xxix) has described the concept of *face* as the "central moment of all of Levinas's phenomenology" (p. xxix) and goes on to explain that while the theoretical structure of his work begins with an "ontological elucidation of what it mean to be existent", his subsequent analysis of the intentional or transcending led him to a region "otherwise than being" (p. xix). This latter reference is to his most notable later work (Lévinas, 1998a) which Moran sees in part as a response to criticism by Jacques Derrida. Levinas's use of the term "face" has become the hallmark of his legacy and in his writing stands for the "real concrete presence" of another which "blossomed into a metaphor" for many aspects of the person and culture (Moran, 2000, p. 347). I will now propose a definition of responsible innovation inspired by Levinas to complement von Schomberg's (2013) definition above (Costello & Donnellan, 2008).

> *Proposition*: An innovation is a human activity resulting in an artifact, idea, practice, organization, learning or information system- perceived to be new by the unit of adoption - that is the cognizant of the Levinasian "Other". Consequently, ethical issues must be considered that affect initiating, implementing and using the innovation together with the associated employment of resources.

Bergo (2007) states that Levinas's "phenomenological descriptions of intersubjective responsibility are built upon an analysis of living in the world" and are unique to him. However, this would seem a little unfair to Husserl, who many years earlier in his lectures on Nature and Spirit had "maintained that an objective external world can only be experienced inter-subjectively (i.e. by a plurality of individual knowing subjects) who are in a position to exchange information with each other" (Posselt, 2005, p. 53).

3.11 Levinas and Heidegger

Now, I will examine the relationship of Levinas with one of the most influential phenomenologists of our time: Martin Heidegger, drawing from a number of studies (Bergo, 2007; C. Davis, 1996; Honderich, 2005; Lingis, 1998; Moran, 2000). The reason I am introducing this connection here is that I believe that Levinas counteracts the more pessimistic views of Heidegger concerning the role of technology. Levinas summed this up in a statement that without technology, "humanity [...] could not be fed" (Lévinas, 1998c) quoted in Bergen (2017). I will develop this argument further in the concluding chapter. To begin, we must admire Levinas's timing: arriving in Freiburg in 1928 where he was able to attend the last lectures of Husserl on phenomenological psychology and intersubjectivity, together with Heidegger's first seminars when he became Husserl's successor. Levinas was very appreciative of the amount of time afforded to him by Husserl but was clearly captivated by what Inwood (2005) describes as the "enthralling brilliance" of Heidegger's lecturing style. During this period, Moran tells us that Levinas sided with Heidegger's formulation of "being-in-the-world" against Husserl's "transcendental idealism". His first publication was a review of Husserl's *Ideas I* where in the conclusion he rejected "quasi-solipsistic egological reduction" as a means to pursue authentic objectivity and presented the first steps in his long journey of phenomenological intersubjectivity (Moran, 2000, p. 323). Significantly, Levinas's award winning thesis presented for his doctorate at the University of Strasburg in 1929 and published the following year was titled *The Theory of Intuition in Husserl's Phenomenology*. Both Davis and Moran describe Levinas's reorientation from his initial use of Heidegger's phenomenology to support his critique of Husserl's subjectivism, to the point where Heidegger became the object of a "complete rebuttal"(C. Davis, 1996, p. 21). Two factors seem to have come into play here. Firstly in the purely philosophical realm where he is quoted by Davis (1996, p. 21) as concluding that the Heideggerian destruction of metaphysics enforces a major flaw in Western thought:

> …this supremacy of the Same over the Other seems to me to be entirely maintained in the philosophy of Heidegger …
>
> He does not destroy, but rather he epitomizes a whole current of Western thought.
>
> The *Dasein* that Heidegger puts in the place of the soul, of consciousness, of the Ego, retains the structure of the Same.

The second important factor was his realisation, when alerted by Alexandre Koyré, of the influence of National Socialism on Heidegger which according to Moran horrified Levinas. Furthermore, it opened his eyes to Heidegger's emphasis on *authenticity* as in fact masking a "self-centred weakness" that was open to exploitation by such totalitarian ideologies.

One aspect of the work of Levinas that also needs to be addressed is literally a matter of semantics and is important for this study. Levinas did not discuss issues such as laws, rules, rights, duties and the language or logic of ethical enquiry (C. Davis, 1996, p. 47). He had a more ambitious programme that of developing a "first philosophy" or what Derrida called writing an "ethics of ethics" (Bergo, 2007). Both Moran and Davis translate this word, from the French *éthique*, as his study of "the ethical" analogous to someone studying "the political" rather than just politics. Vandekerckhove (2019) in his comments on Costello (2019) argues that there is a tricky paradox when relating the work of Levinas to innovation since Levinas' ethics of responsibility has its foundation in the proximity of the Other. He contends that "in innovation there is no proximity: consequences are largely unknown and, in the future" (ibid., p. 211). However, Starks-Estes (2016) points out that Levinas' work has been applied to the analysis of online social networks as well as offline relationships. Proximity in the digital age can be as much virtual as physical. An unfortunate example of this is the reality of cyberbullying in the realm of the Internet.

3.12 Conclusions

This chapter synthesises the area of design and innovation with relevant literature on "responsibility" to facilitate communication and address a major concern of Alasdair MacIntyre (2006, p. 4):

> Imprisoning philosophy within the professionalizations and specializations of an institutionalized curriculum after the manner of our own contemporary European and North American culture is arguably a good deal more effective in neutralizing its effects than either religious censorship or political terror.

I have also attempted in this chapter to respond to MacIntyre's lament that in the case of Edith Stein "there is a general neglect of her work in the English-speaking world" (ibid., p. vii).

However, one question remains that can create unease: Is the Levinasian *Other* overly sentimental and presents us with an untenable perfection? No I believe – Levinas himself answers this in his following poetic style (Lévinas, 1998b, p. 49):

> Autonomy, the philosophy which aims to ensure the freedom, or the identity, of beings, presupposes that freedom itself is sure of its right, is justified without recourse to anything further, is complacent in itself, like Narcissus. When, in the philosophical life that realizes this freedom, there arises a term foreign to the philosophical life, other - the land that supports us and disappoints our efforts, the sky that elevates us and ignores us, the forces of nature that aid us and kill us, things the encumber us or serve us, men who loves us or enslave us- it becomes an obstacle; it has to be surmounted and integrated into this life.

Finally, we should be cognisant of the reason why, according the Moran (2000, p. 180), Husserl embarked on writing *The Crisis of European Sciences*. It was "an attempt to alert the world to the increasing danger of the collapse of the genuinely scientific and philosophical outlook which had marked out the progress of the West since the time of the Greeks".

In the next chapter, I will explore an important but contentious topic for the teaching of design and innovation, namely, the role of theory.

Chapter Summary

- There is an emerging and important discourse on the theme of responsible research and innovation (RRI) in the context of the ubiquity of technology.
- Empathy is a vital early step in the process of design and innovation.
- Philosophy is a necessary ingredient to provide a basis for the teaching of the topic.
- Responsible research and innovation (RRI) is becoming an increasingly important factor in European research policy.
- Edith Stein is the philosopher of empathy, *par excellence*, and her work provides a fruitful underpinning to understand and teach the idea of responsibility.
- The philosophical lens of Emmanuel Levinas and his concept of the "face of the Other" can complement Stein's work in teaching responsible design and innovation.
- Engagement with these debates provides an antidote to the potential pervasiveness of blind technological thinking and tendencies to technological somnambulism.

References

Ainley, A. (2005). Levinas, Emmanuel. In T. Honderich (Ed.), *The Oxford companion to philosophy* (2nd ed., pp. 512–513). Oxford, UK: Oxford University Press.

Ameriks, K. (1977). Husserl's realism. *The Philosophical Review, 86*(4).

Baseheart, M. C. (1989). Foreword to the third edition. In *On the problem of empathy (The collected works of Edith Stein, vol. 3) -translated by Waltraut Stein, Ph.D.* Washington, DC: ICS Publications.

Bergen, J. P. (2017). Responsible innovation in light of Levinas: Rethinking the relation between responsibility and innovation. *Journal of Responsible Innovation, 4*(3), 354–370. https://doi.org/10.1080/23299460.2017.1387510.

Bergo, B. (2007). Emmanuel Levinas. In E. N. Zalta (Ed.), *The Stanford encyclopedia of philosophy* (Fall 2014 ed.). http://plato.stanford.edu/archives/fall2014/entries/levinas/

Blok, V. (2014). Look who's talking: Responsible innovation, the paradox of dialogue and the voice of the other in communication and negotiation processes. *Journal of Responsible Innovation, 1*(4).

Blok, V., Hoffmans, L., & Wubben, E. F. M. (2015). Stakeholder engagement for responsible innovation in the private sector: Critical issues and management practices. *Journal on Chain and Network Science (SPECIAL ISSUE: Responsible Innovation in the Private Sector), 15*(2), 147–164.

Blok, V., & Lemmens, P. (2015). The emerging concept of responsible innovation. Three reasons why it is questionable and calls for a radical transformation of the concept of innovation. In B.-J. Koops, I. Oosterlaken, H. Romijn, T. Swierstra, & J. V. D. Hoven (Eds.), *Responsible innovation 2: Concepts, approaches and applications.* Heidelberg/New York/Dordrecht/London: Springer.

Chesbrough, H. W. (2003). *Open innovation: The new imperative for creating and profiting from technology.* Boston: Harvard Business School.

Costello, G. J. (2015). *Working paper: Review of innovation process frameworks.* Research@THEA Repository. https://research.thea.ie/handle/20.500.12065/361. Accessed July 2019.

Costello, G. J. (2019). The philosophy of innovation in management education: A study utilising Aristotle's concept of Phronesis. *Philosophy of Management, 18*(3), 215–230. https://doi.org/10.1007/s40926-018-00104-7.

Costello, G. J., Cresham, M., & Donnellan, B. (2007). Empathy and teamwork: Reflections on the legacy of Claudio Ciborra through the phenomenology of Edith Stein. *15th European Conference on Information Systems, St. Gallen, Switzerland, June 7–9, 2007.*

Costello, G. J., & Donnellan, B. (2008). Seeking the face of innovation with the ethical compass of Emmanuel Levinas. In G. Leo6n, A. Bernardos, J. Casar, K. Kautz, & J. DeGross (Eds.), *IFIP International Federation for Information Processing, volume 287, open IT-based innovation: Moving towards cooperative IT transfer and knowledge diffusion* (pp. 97–117). Boston: Springer.

Czap, N. V., Czap, H. J., Khachaturyan, M., Lynne, G. D., & Burbach, M. (2012). Walking in the shoes of others: Experimental testing of dual-interest and empathy in environmental choice. *The Journal of Socio-Economics, 41*(5), 642–653.

Darwall, S. (1998). Empathy, sympathy, care. *Philosophical Studies, 89*(2), 261–282.

Davis, C. (1996). *Levinas : An introduction.* Cambridge, UK: Polity Press.

Florea, L., Cheung, Y. H., & Herndon, N. C. (2013). For all good reasons: Role of values in organizational sustainability. *Journal of Business Ethics, 114*, 393–408.

Galbraith, J. K. (1967). *The New Industrial State* (2nd ed.). London, United Kingdom: Pelican Books.

Gordon, R. M. (1999). Empathy. In R. Audi (Ed.), *The Cambridge dictionary of philosophy* (p. 261). New York: Cambridge University Press.

Gorry, G. A., & Westbrook, R. A. (2011). Once more, with feeling: Empathy and technology in customer care. *Business Horizons, 54*, 125–134.

Heidegger, M. (1977). The Question Concerning Technology *The Question Concerning Technology: And Other Essays* (pp. 3–35). New York & London Garland Publishing, Inc.

Heise, U. K. (2004). Science, technology, and postmodernism. In S. Connor (Ed.), *The Cambridge companion to postmodernism*. Cambridge University Press.

Honderich, T. (Ed.). (2005). *The Oxford companion to philosophy* (2nd ed.). Oxford, UK: Oxford University Press.

ICS. (2007). On-line review of "*Edith Stein: An investigation concerning the state translated by Marianne Sawicki*". Available on line through http://www.icspublications.org/

Inwood, M. J. (2005). Heidegger,Martin. In T. Honderich (Ed.), *The Oxford companion to philosophy* (2nd ed., pp. 371–375). Oxford, UK: Oxford University Press.

Klein, M. (2005). Responsibility. In T. Honderich (Ed.), *The Oxford companion to philosophy* (2nd ed., pp. 815–816). Oxford, UK: Oxford University Press.

Kline, S. J. (2002). What is technology? In R. Scharff & V. Dusek (Eds.), *Philosophy of technology: The technological condition*. Oxford, UK: Blackwell.

Leonard, D. A., & Rayport, J. (1997). Spark innovation through empathic design. *Harvard Business Review, 75*(6 November–December), 102–113.

Lévinas, E. (1998a). *Otherwise than being, or, beyond essence*. Pittsburgh, PA: Duquesne University Press.

Lévinas, E. (1998b). Philosophy and the Idea of Infinity. In *Collected philosophical papers* (A. Lingis, Trans.) (pp. 47–59). Pittsburgh, PA: Duquesne University Press.

Lévinas, E. (1998c). Secularization and hunger. *Graduate Faculty Philosophy Journal, 20/21*(2/1), 3–12. https://doi.org/10.5840/gfpj199820/199212/199813.

Lingis, A. (1998). Translators's introduction. In *Collected philosophical papers: Emmanuel Levinas* (A. Lingis, Trans.) (pp. vii–xxxi). Pittsburgh, PA: Duquesne University Press.

MacIntyre, A. C. (2006). *Edith Stein: A philosophical prologue, 1913–1922*. Lanham, Maryland: Rowman & Littlefield.

Makkreel, R. A. (1999). Einfühlung. In R. Audi (Ed.), *The Cambridge dictionary of philosophy* (pp. 257–258). New York: Cambridge University Press.

Moran, D. (2000). *Introduction to phenomenology*. London/New York: Routledge.

Niewiadomski, R., Ochs, M., & Pelachaud, C. (2008). Expressions of empathy in ECAs. In H. Prendinger, J. Lester, & M. Ishizuka (Eds.), *Intelligent virtual agents volume 5208 of the series lecture notes in computer science* (pp. 37–44). Berlin Heidelberg: Springer.

Nota, J. (1988). Edith Stein and Martin Heidegger. In *Carmelite studies* (pp. 50–73).

ODE. (2006a). Technology. In *Oxford Dictionary of English* (2nd ed., Revised). Oxford, UK: Oxford University Press.

ODE (Ed.). (2006b). *Oxford Dictionary of English* (2nd ed., Revised). Oxford, UK: Oxford University Press.

Pettersson, J. S. (2002). Visualising interactive graphics design for testing with users. *Digital Creativity, 13*(2), 144–156.

Pitt, L. F., Watson, R. T., & Kavan, C. B. (1995). Service quality: A measure of information systems effectiveness. *MIS Quarterly, 19*(2), 173–187.

Posselt, T. R. (2005). Edith Stein: The life of a philosopher and carmelite. In S. M. Batzdorff, J. Koeppel & J. Sullivan (Eds.). Washington, DC: ICS Publications.

Potapova, A., & Rodionov, S. (2014). Universal empathy and ethical bias for artificial general intelligence. *Journal of Experimental & Theoretical Artificial Intelligence, 26*(3), 405–416.

Pučėtaitė, R., Novelskaitė, A., Lämsä, A., & Riivari, E. (2016). The relationship between ethical organisational culture and organisational innovativeness: Comparison of findings from Finland and Lithuania. *Journal of Business Ethics, 139*(4), 685–700. https://doi.org/10.1007/s10551-10016-13051-10558.

Owen, R., Heintz, M., & Bessant, J. (Eds.). (2013). *Responsible innovation: Managing the responsible emergence of science and technology in society*. London: Wiley.

Richmond, S. (2005). Empathy. In T. Honderich (Ed.), *The Oxford companion to philosophy* (2nd ed.). Oxford, UK: Oxford University Press.

Riivari, E., Lämsä, A.-M., Kujala, J., & Heiskanen, E. (2012). The ethical culture of organisations and organisational innovativeness. *European Journal of Innovation Management, 15*(3), 310–331.

Riivari, E., & Lämsä, A. M. (2014). Does it pay to be ethical? Examining the relationship between organisations' ethical culture and innovativeness. *Journal of Business Ethics, 124*(1), 1–17.

Sand, M. (2018). The virtues and vices of innovators. [journal article]. *Philosophy of Management, 17*(1), 79–95. https://doi.org/10.1007/s40926-017-0055-0.

Sandler, R. L. (2013). Introduction: Technology and ethics. In R. L. Sandler (Ed.), *Ethics and emerging technologies*. Basingstoke, UK: Palgrave Macmillan.

Sawicki, M. (2000). Editor's introduction. In M. Sawicki & M.-C. Baseheart (Eds., Trans.), *Philosophy of psychology and the humanities, the collected works of Edith Stein* (pp. xi–xxiii). Washington, DC: ICS Publications.

Starks-Estes, L. S. (2016). From face to Facebook: Levinas's radical ethics and "Shakespeare Friends". *Borrowers and Lenders: The Journal of Shakespeare and Appropriation, 10*(1). http://www.borrowers.uga.edu/1802/show

Stein, E. (1989). *On the problem of empathy (The collected works of Edith Stein; vol. 3) -translated by Waltraut Stein, Ph.D.* Washington, DC: ICS Publications.

Stein, E. (1993). Knowledge, truth, being: Translated by Walter Redmond (The collected works of Edith Stein; vol. 8). In L. Gelber & M. Linssen (Eds.), *Knowledge and faith*. Washington, DC: ICS Publications.

Stein, E. (1996). *Essays on woman* (F. M. Oben, Trans.) (2nd ed.). Washington, DC: ICS Publications.

Stein, E. (2000). *Philosophy of psychology and the humanities*. Washington, DC: ICS Publications.

Stein, W. (1989). Translators introduction. In *On the Problem of Empathy (The collected works of Edith Stein; vol. 3) -translated by Waltraut Stein, Ph.D.* Washington, DC: ICS Publications.

Stilgoe, J. (2013). Foreword: Why responsible innovation? In R. Owen, M. Heintz, & J. Bessant (Eds.), *Responsible innovation: Managing the responsible emergence of science and technology in society*. London: Wiley.

Teresia de Spiritu Sancto. (1952). *Edith Stein (translated from the German by Cecily Hastings and Donald Nicholl)*. London/New York: Sheed & Ward.

Vandekerckhove, W. (2019). Practical wisdom, respect and metaphysics: A broad spectrum for philosophy of management. *Philosophy of Management, 18*(3), 211–214. https://doi.org/10.1007/s40926-019-00123-y.

von Schomberg, R. (2013). A vision of responsible innovation. In R. Owen, M. Heintz, & J. Bessant (Eds.), *Responsible innovation*. London: Wiley.

VTI. (2019). *The Department of Values, Technology & Innovation (VTI). Delft University of Technology*. https://www.tudelft.nl/en/tpm/about-the-faculty/departments/values-technology-and-innovation/. Accessed November 2019.

Womack, J. P., & Jones, D. T. (2003). *Lean thinking: Banish waste and create wealth in your corporation*. London: Free Press.

Zakaras, A. (2003). Isaiah Berlin's Cosmopolitan Ethics. *Political Theory, 32*(4), 495–518.

Chapter 4
The Principle of Theory

4.1 Introduction

While engaged in my studies, I was faced with the daunting task of finding a suitable theory for my work. I began searching for an answer to the question: What exactly is a theory? Eventually I confided my confusion to a senior academic who had written on the subject. He calmed me down by assuring me that he would send me a seminal paper on the topic. I waited expectantly, and finally the paper appeared in my inbox. When I saw the heading of the paper, my heart sank – it was entitled "What theory is not"! (Sutton & Staw, 1995). After several years examining the subject, I now concur that it is easier to say what a theory "is not" than what a theory "is". This chapter is the result of my pursuit of theories to help our understanding of design and innovation in order to support the teaching of the subject.

The chapter proceeds as follows. First, I argue for the importance of an innovation-based view (IBV) of an organisation (Costello, 2018). In doing so, it builds on the current incumbent theories, resource-based view (RBV), knowledge-based view (KBV) and dynamic capabilities in light of recent developments in the academic literature and organisations. Second, I will draw on the work of Carol Slappendel, who contends that design and innovation is an interactive process, and I will suggest that this notion should be incorporated in any theoretical framework. Third, I will propose that the complexities of design and innovation require an ecological approach by presenting an ecological systems theory (EST). Finally, I will combine these three aspects into a metatheory. In the course of this chapter, the focus of theory development will refer to *innovation* due to its prevalence in the literature. However, such theoretical development applies to both *design and innovation* given the intertwining of both. But before this, I will present some thoughts on what is theory; what is a good theory; and how theory can be conceived using the approach of abduction.

G. J. Costello, *The Teaching of Design and Innovation*, Contemporary Issues in Technology Education, https://doi.org/10.1007/978-3-030-41380-4_4

4.2 Theory Development

In research across almost every discipline, there is a diaspora of underlying theories, along with many applications, implementations and adaptations of those theories. Ruse (2005) defines theory as an "attempt to bind together in a systematic fashion the knowledge that one has of some particular aspect of the world experience". The aim of such endeavour is to gain understanding in order, for example, to explain, predict or describe. Theory building is important, according to Wacker (1998, p. 361), because it "provides a framework for analysis, facilitates the efficient development of the field, and is needed for the applicability to practical real world problems". *HowStuffWorks* provides answers and explanations of how the world works, but I view theory as addressing: *Why stuff works*? Kurt Lewin is credited with the aphorism that "there is nothing as practical as a good theory". However, it is important to determine what constitutes a "good" theory and what distinguishes it from one of lesser value. To this end, I will use the five constructs proposed by Conboy (2009, p. 330) when he examined the development of theory in the area of agile software methods.

Clarity: One of the most fundamental attributes of a concept is that it is clearly communicated and understandable (Dubin, 1976, 1978; Metcalfe, 2004; Weick, 1989). Design and innovation are multifaceted, and the concepts are used in many ways in the literature. For example, in management information systems, inconsistency and contradiction evoke emotive descriptions such as "fragmented adhocracy" (Banville & Landry, 1989) and "crucial but confused" (Checkland & Holwell, 1998).

Theoretical Glue: Behind any good concept or theory, there should be a strong underlying logic and rationale. Whetten (1989) refers to this as a "theoretical glue" which should bind all of the factors together (Conboy, 2009, p. 330). However the analysis of the term innovation has revealed multitude of underlying constructs (Costello & Donnellan, 2008). Consequently, providing a definition and essential rationale for the topic of innovation was a fundamental task before progressing to a theory. Additionally, the final framework will show the theories "bonded" in a nested Russian doll formation.

Cumulative Tradition: A good concept or theory should cumulatively build on existing research (Dubin, 1978), yet this is something that, for example, management researchers have not done particularly well (I. Benbasat & R. Zmud, 1999; Keen, 1980, 1991). Keen (1991) notes that most concepts and areas of concern in research are not as "new" as often claimed and often "turn out to have long roots". The literature on design and innovation has a significant history of theoretical work, but there is much debate on the value of its contribution (Fagerberg, 2005; Wolfe, 1994).

Parsimony: Authoritative works on concept development usually advocate a parsimonious approach, removing any factors which provide little additional value to our understanding (Whetten, 1989). Similar to agile research (Conboy,

2009, p. 331), there is also much redundancy and duplication in the design and innovation literature together with a lot of non-value-added content. A more cohesive approach towards a theory of how design and innovation influence the success of an organisation can provide a reference to help avoid duplication of effort in future research.

Applicability: The range of applications of a concept is a key criterion for judging its quality (Dubin, 1976, 1978; Metcalfe, 2004; Weick, 1989), and so any "good" theory should be applicable in a wide variety of contexts. Theories of the firm, despite criticism and drawbacks, have exhibited a wide range of applicability to the various organisational manifestations. The ubiquity of innovation as a conceptual framework is applicable to large multi-national firms, subsidiaries, small- and medium-sized enterprises, public sector bodies, as well as the important emerging not-for-profit sector, which is a recent and important development in the literature.

Now I will discuss the genesis of theory based on work chiefly gleaned from the philosophical literature. This will involve the concept of "abduction" which is much less well known than "deduction" and "induction", but which is very relevant to the topic of developing theory.

4.3 Abduction

In his 1883 work "Theory of Probable Inference", Charles Sanders Peirce argued that there are three fundamental kinds of reasoning: deduction, induction and abduction (Peirce, 2010, p. 439). Peirce was originally a Harvard chemistry graduate, and his greatest philosophical influence was Kant (Hookway, 2005). Van de Ven (2007) builds on the work of Peirce and proposes that theory building involves three related activities (p. 101):

1. Conceiving or creating a theory (abduction)
2. Constructing or elaborating the theory (deduction)
3. Justifying or evaluating a theory (induction)

Elsewhere Scruton (2004) expresses this taxonomy as follows: the passage down from theory to observation is one of deduction, the passage up from observation to theory is one of induction while abduction denotes the process of forming a hypothesis (p. 190). Hookway (2005, p. 687) describes the logic of abduction as the logic of discovery as it studies how we are guided to a new hypothesis. Also, it is sometimes referred to as *inference to the best explanation* where inference can be explained as an upgrading of belief from existing beliefs due to the gaining of new relevant information. Suppe (1999, p. 1) describes abduction as the canons of reasoning for the discovery, as opposed to the justification, of scientific hypotheses or theories. Van de Ven (2007, p. 140) describes abduction as follows:

> Abduction is a creative form of reasoning that is triggered by encountering anomalies and ends by selecting a plausible or coherent solution that might resolve the anomaly

As Svennevig (2001, p. 1) points out, "practical scientific programs cannot be based on either pure deduction or pure induction. Central to any scientific process is the inferential step from some initial puzzling fact to some theoretical hypothesis which can explain it".

4.4 Chamberlin's Multiple Working Hypotheses

At this stage I will address the question: Does a new theory usurp previous theories? To answer this, I will call on a seminal paper by Thomas C. Chamberlin (1890) (sic) written in the very stylised prose of that era, which continues to be influential. Chamberlin's introduces his method of multiple working hypotheses by dividing research into two broad categories. He calls the initial path secondary, imitative or acquisitive study, which involves closely following the process developed by previous scholars. The second category is described as "primary or creative" study where the approach is to "think independently, or at least individually, in the endeavour to discover new truth". Furthermore, the latter is the most promising avenue for research in subjects "which, while much is known, more remains to be known". Chamberlin then delineates three intellectual methods that have driven the progress of science: the method of ruling theory, the method of the working hypothesis and the method of multiple working hypotheses. He argues convincingly for the universal adoption of the last method in his paper. The dominance of "affection for the favoured theory" causes a narrowing of vision and possibilities, which results in the stilting of research and errors such as misinterpretations and misjudgements. He expresses his concern for the ensuing detrimental research cycle: an initial precipitate explanation gradually evolves into a tentative theory, becomes an adopted theory and finally reaches the dubious status of the ruling theory. Evoking analogies not often found in contemporary academic literature, he points out that research – like love – can suffer from the tendency to blindness. Chamberlin then turns his sharp critical gaze to "the" scientific method of his day: the working hypothesis. He concludes that while it is an improvement, its basic flaw is that it is "but the means for the more ready development of facts and of their relations" and can quickly become the usurper to the title of the ruling theory. Chamberlin proposes that, faced with the complexities of research, the method of multiple working hypotheses provides an antidote to presumptions "that this or that has been done because it accords with our ruling hypothesis or our working hypothesis". The hypothesis of the innovation-based view (IBV), described in the next section, fits within Chamberlin's typology, as while much is known about the topic of design and innovation, much more work is required. Furthermore, IBV demonstrates "independent thinking" and together with other incumbent theories provides researchers with a toolkit of multiple working hypotheses to address the complexities of design and innovation research.

I will now address the theory of design and innovation in three ways. First, I will argue for the importance of an innovation-based view (IBV) of an organisation.

Second, I will draw on the work of Carol Slappendel who contended that design and innovation is an interactive process. Third, I will propose that the complexities of the subject require an ecological approach by presenting an ecological systems theory (EST) of design and innovation. Finally, I will combine these three aspects into a metatheory.

4.5 Theory 1: An Innovation-Based View

This section argues that the success of a modern organisation stems from its ability to harness and utilise its design and innovation potential, which is increasingly exercised in an open environment. This view postulates the need to shift attention from resources and knowledge capabilities and focus more on design and innovation capabilities. Now I will initially provide a brief overview of three theories traditionally associated with the "firm" but which I will now apply to the "organisation" which is more in keeping with modern developments in practice and in the literature. The concepts of resources, knowledge and innovation also apply to the not-for-profit sector, the public sector and the ever-growing services sector.

Theories of the firm are conceptualisations and models of business enterprises which explain and predict their structures and behaviours (Grant, 1996). "As a result, there are many theories of the firm which both compete in offering rival explanations of the same phenomena and complement one another in clarifying these different phenomena" (ibid., p. 109). The movement away from viewing an organisation as a "black box" is usually traced to the work of Edith Penrose who focused on the internal resources of a firm especially with respect to the management and experience it acquires over time (Sautet, 2000). Penrose, working in the 1950s, argued that the growth of a firm is not just the response to market changes but exhibits an internal developmental process (Penrose, 2009). This chapter, while respecting economics scholarship, is in the tradition of Penrose which was further developed in the strategic management literature (Hoskisson, Hitt, Wan, & Yiu, 1999). I will broadly refer to this lens as the managerial view incorporating the resource-based view (RBV), knowledge-based view (KBV) and dynamic capabilities.

The RBV has been the subject of much academic debate and no little criticism over the past 20 years. It has become one of the most influential and cited theories in the area of management, and it aims to "explain the internal sources of a firm's sustained competitive advantage" (Kraaijenbrink, Spender, & Groen, 2010, p. 2). The RBV perceives the firm as a unique bundle of idiosyncratic resources and capabilities where the primary task of management is to maximise value through the optimal deployment of existing resources and capabilities while developing the firm's resource base for the future (Barney, 1991).

The KBV of an organisation is an outgrowth of the RBV of the firm. The KBV would argue that maximising the value deriving from the firm's resources builds from superior knowledge. In the KBV, the organisation is conceptualised as an institution for integrating knowledge (Grant, 1996). The KBV posits that knowledge is the most strategically important of the organisation's resources. With knowledge, organisations become more efficient and effective with the use of scarce resources. This view is gaining increasing attention due to the rapid movements towards knowledge-based societies. However, the danger inherent with the concept of organisational knowledge is that the organisational processes through which individuals engage in these activities may be obscured, by their viewing the organisation as an entity that creates, stores and deploys knowledge (Coakes, 2000).

Another development of RBV and KBV is the theory of dynamic capabilities described by Douma and Schreuder (2008, p. 207) as "the capacity of an organization to purposefully create, extend or modify its resource base". In their seminal paper on the theory, Teece, Pisano and Shuen (1997, p. 516) define dynamic capability as the "ability to integrate, build, and reconfigure internal and external competences to address rapidly changing environments". Dynamic capabilities differ from operational capabilities as they are focused on change and in doing so alter the resource base of an organisation.

4.5.1 Linking Knowledge to Design and Innovation

Theories adhering to the KBV tend to emphasise the ensuing advantages to organisations by way of knowledge acquisition and sharing activities. But such an emphasis is deficient. Innovative outcomes can only materialise when knowledge is utilised. In today's digitally enhanced world, acquiring relevant information can potentially be only a few clicks away. Where once the challenge was accessing and sharing information, the main challenge is now the utilisation of acquired knowledge. Hence, I call for an outlook where the realisation of knowledge in a design and innovation artefact is the core performance differentiator. A case in point is the Whelan, Parise, Valk, and Aalbers (2011) study of why companies fail at open innovation. The companies studied had well developed formal and informal mechanisms to acquire state-of-the-art knowledge from the outside world. But what they failed to pay attention to is how that knowledge would reach the people in the firm with the skills to utilise it. Very often, valuable knowledge imported into the firm remained in storage, never to be used.

Leonard (1998) links knowledge to innovation and claims that the successful innovators are organisations that build and manage knowledge effectively. They are the most enthusiastic about pursuing knowledge and the most likely to harness the power of innovation. Leonard illustrates the dimensions of the core capabilities along which all organisations must innovate: physical systems, managerial systems, skills and norms of behaviours. She views organisations as sites of learning and information transfer rather than physical sites or financial entities. The workforce of a forward-looking organisation must be able to process and manipulate knowledge as well as perform skills.

I will now propose a definition of IBV in the following proposition:

> An innovation-based view considers the ability to ethically manage and use internal and external innovation as the most strategically significant resource of an organisation.

Furthermore, the IBV will be examined using the five constructs proposed by Conboy (2009) together with Peirce's concept of abduction enveloped within an ethical framework. These ideas were introduced above and used as guiding principles for the development of the theory. The constructs used in this theory building process are assessed in Table 4.1.

Now I will develop the discussion of design and innovation beyond a "view" of the organisation by examining its conceptualisation as interactive process within the organisation.

Table 4.1 Building an IBV of an organisation

Construct	IBV taxonomy
Clarity	The term design and the term innovation are multifaceted and even contentious in the voluminous and eclectic literature base. Providing an all-encompassing classification of the concept is indeed a lofty undertaking. However, a definition is offered based on the author's analysis of a significant number of antecedent definitions from the literature in order to clarify the term as applied to the IBV
Theoretical glue	The underlying logic and rationale for the IBV was compiled from a structured examination of the notion of innovation vis-à-vis incumbent theories as well as recent developments in the nature of the organisations largely driven by the rapid expansion of ICT. Hence the proposed taxonomy of IBV was *glued* together from the exploration of the concepts of design, *innovation*, *theory* and *nature of the organisation*
Cumulative tradition	The cumulative tradition of scholarship on design and innovation has its origins in the works of Schumpeter and Simon and has been significantly developed over the past number of decades. IBV is placed in the development of the managerial theories originating from scholars such as Edith Penrose
Parsimony	The definition of innovation was developed from an investigation and parsing of many antecedent definitions of innovation. There was much duplication and redundancy in the literature main concepts uncovered during the analysis. Consequently, the IBV proposed in the study is a concise formulation that facilitates future investigation and empirical studies
Applicability	The development in ICT and the nature of organisations has resulted in the incumbent theories having limited applicability in the current business landscape. As a result, the encompassing nature of the concept of innovation has ensured that IBV is valid in the broad spectrum of organisational situations. For example, the IBV includes the emerging not-for-profit sector
Abduction	IBV makes an academic contribution by providing an example of initiating theory using abduction (conceiving of theory) which has recently been championed by Andrew Van de Ven. Antecedent theories were largely developed, analysed and researched using the traditional concepts of induction and deduction
Ethics	Ethics is a fundamental area of philosophical inquiry and of managerial practice. The inclusion of ethics in the IBV was a response to the virtual absence of references to this topic in an analysis of innovation definitions in the literature. It is proposed that this approach in the current study is an important contribution to the investigation of the nature of an organisation

4.6 Theory 2: An Interactive Process

Carol Slappendel (1996) classified the innovation literature in terms of three theoretical perspectives based on the antecedent work of Pierce and Delbecq (1977). These three theoretical approaches, the individualist perspective, the structuralist perspective and the interactive process perspective, are presented in the sequence that reflects their historical development. She argued that the increasing growth in innovation publications necessitates that both researchers and students "establish mental models of the domain", which is one of the main objectives of this chapter. The first category in her cognitive map is that of the individualist perspective which is characterised by the basic assumption that individuals cause innovation. Researchers holding this view propose that certain individuals have personal qualities or "traits" which "predispose them to innovative behaviour" and that they make rational decisions based on the economic concept of "utility". Studies of this type use concepts such as "age, sex, education level, values, personality, goals, creativity and cognitive style" to examine innovation. While there is general acceptance of the importance of individual roles such as champions, leaders and entrepreneurs in the innovation process, this perspective has been firstly criticised for its unreasonable assumption that innovative decisions are made by autonomous agents and secondly due to evidence that suggests individual characteristics can be subsumed by organisational roles and positions. She quoted studies, by among others Saren (1987) and Van de Ven (1986), which argued that the complex, nonroutine and often irrational characteristics of the innovation process require the assistance of other individuals in an organisational context. The former author also made the salient point that it is just as important to understand why some people resist innovative activity as why some people support innovation.

The second category, that of the structuralist perspective, proposes that innovation is determined by the structural characteristics of the organisation. The proponents are from several different theoretical schools which according to Astley and Van de Ven (1983) "share a common deterministic orientation by which organizational behavior is seen to be shaped by a series of impersonal mechanisms that acts as external constraints on actors". The shift to this perspective coincided, according to Slappendel, with the increased promotion of "positivist epistemology in business research" and the availability of increasing computing power for "quantitative data analysis". However, the main criticism of the approach is the organisational features such as technology and strategy tend to be "reified" and treated as objective realities per se. Furthermore, Slappendel suggests that the widespread one-dimensional use of "cross-sectional questionnaire surveys" fails to capture sociopolitical influencers such as government policy. For example, many studies agree that the uncertainty generated by rapid environmental change stimulates innovation and that innovation is facilitated where an organisation has extensive communication channels with actors in its environment. She concluded that the individualist and structuralist perspectives tend to facilitate research projects that focus on identifying chief determinants of innovation and which employ cross-sectional surveys. Another interesting point was that many researchers seemed to gravitate to either of these perspectives more for pragmatic reasons, such as restricting the scope of the research project, than from any ideological conviction of their incompatibility. In the final analysis, Slappendel proposes that adopting either of these "monistic" approaches will "inevitably lead to simplistic theorizing of complex phenomena".

The development of the third approach, the interactive process perspective, resulted from a reaction by scholars to the linear "stage-to-stage" notion of the innovation process and from calls that researchers view innovation as a dynamic process in a continuously changing environment. This evolution in thinking requires that any attempt at the generation of theory should address "the complex and paradoxical relationship between action and structure" over time. It also needs to endeavour reconciling both individual and structuralist consideration by analysing their interconnection. One study of interest is that by Walton (1987), who proposed a framework that emphasised the interaction of the factors which take into consideration individual, organisational and environmental features. Walton acknowledges his debt to the work of Pettigrew (1987) who argued that change should be analysed in terms of the dynamic interplay between context, content and process with the preferred research approach of "a historical method involving longitudinal case studies". Pettigrew's work, while primarily focusing on strategic change, has informed and influenced research on strategic innovation. Slappendel's analysis of these studies illustrates the following important aspects of the third perspective: the rejection of the "rational economic model of decision making" with the associated attention to the political context; the emphasis on "understanding the dynamic nature of the innovation process"; the belief that innovations may be transformed by the process itself; and finally the methodological implications which resulted in the prevalence of longitudinal case studies with a focus on induction and in some situations the use of grounded theory. She agrees with the researchers who have rejected the conceptualisation of innovations as isolated, static objects or practices and that have argued for the analysis of complex innovation configurations and changes of innovation features over time. This opportunity for developing new conceptualisations is an important motivation for this book. However, several theoretical and methodological challenges face those who undertake researching the interconnectivities of innovation. These include the tendency to lapse back into the "description of uni-directional models"; the lack of training and knowledge about the approach; the cognitive limitations of a researcher to process the large amount of information coming from a fast-changing environment; and, finally, the linguistic constraints of the words used to describe the research. One proposal to deal with some of these challenges is to involve pairs or teams of researchers. The suitability of case study research to generate new insights into complex organisational phenomena suggests a strong methodological compatibility with the interactive process perspective notwithstanding the ongoing debate on generalisation (Yin, 1994) and the time-consuming nature of the approach. I have captured Slappendel's perspectives in a diagrammatic form in Fig. 4.1.

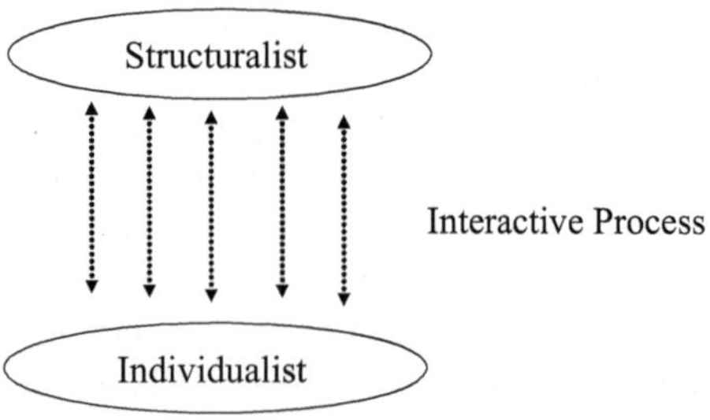

Fig. 4.1 A diagrammatic view of Slappendel's perspectives

I agree with Slappendel's important analysis of innovation, especially her contention that due to the increasing level of organisation change, the interactive process perceptive will become more established. However, I believe that several factors call for an enhanced theoretical framework. The first point concerns the importance of the environment on the innovation process which was raised in her reference to Walton's work. Recent studies such as that by Crowston and Myers (2004) have proposed a broadening of research from its present focus on individual and organisational aspects to industry level that should include an economic perspective, an institutional perspective and a sociocultural perspective. The second point is that Slappendel's perspectives do not consider the multilevel discontinuities driven by the rapid advances of information and communication technology (ICT) which have taken place since the publication of her paper, chiefly through the development of the Internet. These paradigmatic shifts have caused fragmentation of organisational boundaries, resulting in a move towards open and user-led innovation (Chesbrough, 2003; von Hippel, 2005) and the development of social networking and networks of practice (Whelan, 2007). To address these concerns, I will now propose a higher-level ecological systems theory (Costello, 2010).

4.7 Theory 3: An Ecological System

In this section I will introduce ecological systems theory and argue that it provides a suitable framework for researchers to approach the topic of design and innovation. The argument builds on the above perspectives of IBV and of Slappendel (1996) and takes into account important developments in design and innovation since the publication of her work.

4.7.1 An Introduction to Ecological Systems Theory

Urie Bronfenbrenner spent most of his professional career as Professor of Human Development, Family Studies and Psychology at Cornell University. His development of ecological systems theory (Bronfenbrenner, 1979) is regarded as having revolutionised studies in these areas by shattering barriers and building bridges among the social science disciplines. Previous to Bronfenbrenner's work, the study of human development was compartmentalised among psychology, sociology, anthropology, economics and political science. However, through the concept of the ecology of human development, these disparate environments were integrated into a holistic conceptual framework of interdependent nested systems where human development was viewed as a continuum (Lang, 2005). Bronfenbrenner viewed a "child's development within the context of the system of relationships that form his or her environment" with each complex "layer" influencing the development (Paquette & Ryan, 2001). His own conception of the theory was as "a set of nested structures, each inside the next, like a set of Russian dolls"(Bronfenbrenner, 1979, p. 3). He acknowledges the debt he owes to the theories of Kurt Lewin who expressed behaviour as a function "of the interplay between person and environment". This contention also dovetails with Bessant's (2003) conclusion that in dealing with the challenges of innovation, the "key management task lies in creating and reinforcing patterns of behavior".

Cranefield and Yoong (2007), building on the work of McLeroy, Bibeau, Steckler, and Glanz (1988), have argued that ecological systems theory can contribute to debate on "relevance" by delivering an "enriched understanding of the domain of IS practice". Following these authors, I will firstly describe each nested layer of a modified Bronfenbrenner model, shown in Fig. 4.2, where the "patterned behaviour" is determined by the following:

1. Individual level: intrapersonal factors – characteristics such as knowledge, attitudes, behaviour, self-concept, skills, etc. The developmental history of the person is also included.
2. Microsystem: interpersonal processes and primary groups – formal and informal social network and social support systems, including the family, work group and friendship networks.
3. Mesosystem: institutional factors – social institutions with organisational characteristics, with formal (and informal) rules and regulations for operation.

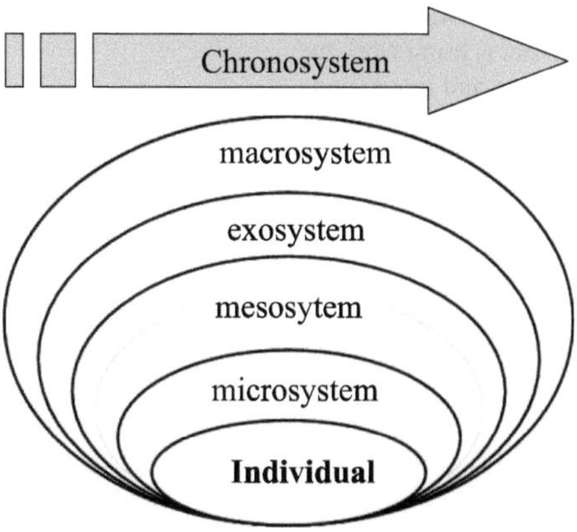

Fig. 4.2 Ecological systems framework. (Adapted from Cranefield and Yoong (2007))

4. Exosystem: community factors – relationships among organisations, institutions and informal networks within defined boundaries.
5. Macrosystem: public policy – local, state and national laws and policies.
6. Chronosystem: this was a later addition by Bronfenbrenner (2004) and was not taken into account by McLeroy et al. This concept incorporates change or constancy over time not only in the characteristics of the person but also of the environment in which that person lives and works.

Having presented the ecological systems theory, I argue that it should not simply to be viewed as an isolated attempt to impose a novel framework on the area of design and innovation but builds on and extends the theoretical perspectives proposed by Carol Slappendel that we referred to earlier.

4.7.2 An Ecological Systems Framework for Design and Innovation

Based on the foregoing analysis, I will now present a framework to analyse design and innovation based on Bronfenbrenner's theory. The structure is based on the implicit assumption that design and innovation originates in the human "person" but is significantly influenced by interaction and interconnection with the five other layers. The framework is illustrated in Fig. 4.3.

The revised innovation framework is now described, and a small number of references are included for the purpose of illustration.

1. *Personal Dimension*: this layer includes the intrapersonal characteristics that assist or inhibit the aptitude to design and innovate. Development of knowledge, skills and competencies through education and training to support design and innovation both in terms of creative invention and of implementation are relevant here (Amabile, Hadley, & Kramer, 2003). For philosophical reasons,

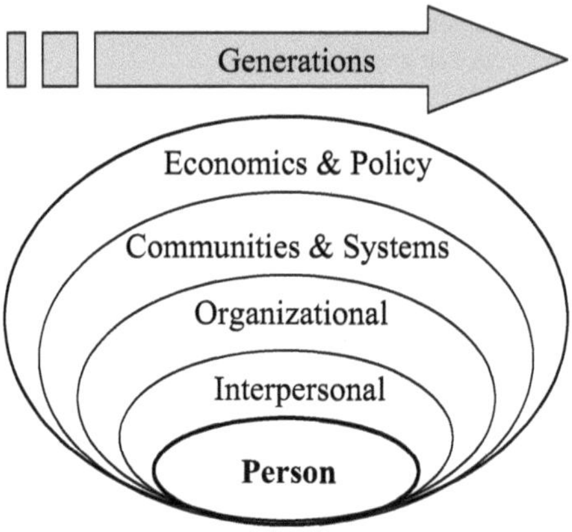

Fig. 4.3 An ecological systems framework for design and innovation

I have replaced the term "individual" as we believe that the origin of design and innovation is the "acting person" (see Chap. 2).

2. *Interpersonal*: formally this dimension will include the ability to contribute to and direct teams or work groups. Informally it will include social networks, communities of practice and personal contacts, both inside and outside the organisation. Interpersonal attributes such as empathy will also be deemed relevant in this layer (E. Stein, 1989).
3. *Organisational*: the characteristics of the organisation that the person is a member of will be significant for this layer. Culture, climate and the management of design and innovation with its associated change will influence the person's tendency to design and innovate (Goffin & Mitchell, 2005).
4. *Communities and Systems*: this layer will include relationship of the organisation with peer organisations, academic institutions and state-sponsored support bodies (Etzkowitz & Leydesdorff, 2000). The layer will also encompass formal and informal networks, clusters that support design and innovation, National Systems of Innovation (NSI) (Lundvall, 1995) and the area of inter-organisational systems (IOS) which is having increasing influence on business to business (B2B) and business to government relationships.
5. *Economics and Policy*: this dimension will include design and innovation policy of local, regional, state (EGFSN, 2017), and supranational (e.g. the European Union) indicators of innovation (OECD, 2005) and important economic theories of innovation (Schumpeter, 1934).
6. *Chronological Generations*: analogous to human development, "generations" can encompass several concepts. At a macro level, it will take cognisance of the time dimension of the design and innovation environment which has been, for example, outlined in Rothwell's (1994) taxonomy of innovation processes. At the organisational level, this would involve assessing the maturity level such as the "archetypes" of innovation proposed by Tidd, Bessant, and Pavitt (2005).

This section has proposed three aspects of design and innovation theory: as a view; as an interaction; and as an ecological system. Now I will incorporate these three aspects into a metatheory.

4.8 Metatheory

The work of Andrew Van de Ven has made a significant contribution to innovation scholarship since the early 1980s. This pioneering work was carried out during the Minnesota Innovation Research Program (MIRP), and its publications are generally known as the Minnesota studies (Andrew H. Van de Ven, Angle, & Poole, 2000). MIRP is regarded as one of the most significant studies on how organisations initiate, develop and implement innovative products and services. The MIRP involved a large number of researchers carrying out longitudinal studies of diverse innovations, over an extended period of time and across a number of disparate organisations. One of the most important results that emerged was the need for a metatheory.

Poole and Van de Ven (2000, p. 638) propose that a satisfactory metatheory must meet the following conditions:

1. Incorporate models of both global (macro and long-run) and local (micro and short-run) development
2. Precisely specified the motor driving development at both levels
3. Spell out inter-level relationships

The ideal metatheory should provide "a conceptual map" in order to explain how innovations "develop, grow, and terminate over time". One major flaw identified in existing theoretical frameworks was the lack of connection between local phenomena and the wider global context of innovation processes. Their following description will be quoted in full as it is extremely pertinent to this work:

> Consideration of local and global models leads one to examine the relations between innovation processes at different levels of analysis, such as how individuals relate to project teams, teams to organizations, organizations to a larger industry community.

The importance of the metatheory being able to accommodate the dynamic nature of the design and innovation process is stressed. Furthermore, any proposed solution should provide guidance on conditions for switching between subsidiary theories and models. Their statement that "previous work has largely ignored this global/local distinction" is very direct. Poole and Van de Ven then describe in more detail what they mean by the term global and local in this context. Global models take into account innovation characteristics "such as economic trends, social needs, the legal system, cultural norms, and the long-term institutional arrangements". On the other hand, local models deal with more immediate factors such as "motivation level and group interaction processes, as well as direct macro influences such as organization structure, resource control, and competition". They summarise this conceptualisation using an analogy.

> [The] worldview in global models tends to be that of an astronaut in orbit, while it is that of the person-on-the-street in local models.

Furthermore, they attribute the lack of global/local connection to the immaturity of developmental theories which should resonate later in our description of the human developmental model. They then go on to argue for the central place of the individual in any general theory of innovation. A main finding of the MIRP studies is that it would be erroneous to consider the role of the individual as a rather mechanistic output of the process of innovation and other "impersonal factors". However, the important role of key individuals did not emerge in isolation but as intimately associated with the organisational context. Successful innovation is "premised on building an organization that can nurture the idea, garner resources, overcome obstacles, and orchestrate development". In addition, the development of innovations was found to be influenced by "key external resource controllers", and "the institutional context" was, in some cases, crucial. However, the relationship between the innovation and its environment should not be considered only as a one-way street. One Minnesota case study described how a revolutionary medical device innovation, a

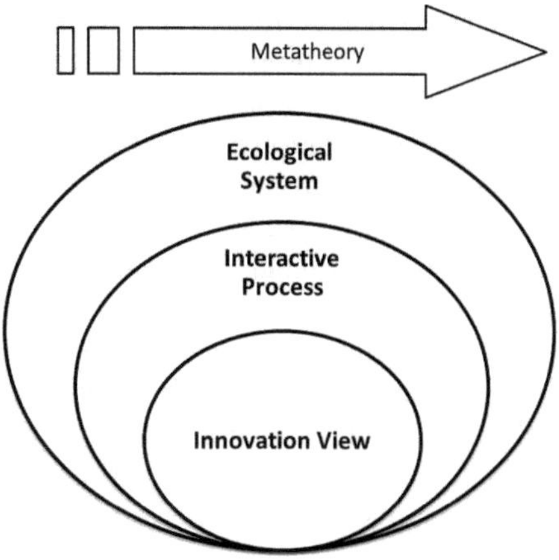

Fig. 4.4 A metatheory of design and innovation

cochlear transplant, had spawned a new industrial sector. Poole and Van de Ven went on to outline in more detail their tentative formulation of a nascent metatheory. Their justification was based on the conclusion that a "single theory cannot encompass the complexity and diversity of process patterns observed across the MIRP innovation studies". Furthermore, they predicted that a "truly general typology and theory of innovation will be a long time coming" but hoped that their paper would "stimulate new patterns of reasoning that may promote further developments in building theories of theories". This chapter is a response to their challenge, and the proposed metatheoretical framework is shown in the conceptualisation provided in Fig. 4.4. Chapters 6 and 7 will explore how this metatheoretical framework can be incorporated into the teaching of design and innovation.

4.9 Conclusions

Gregor (2006, p. 613) contends that "thinking clearly about the nature of theory" has importance for both research and practice and that leading journals expect papers to have a strong theoretical foundation. In the same essay, she classified metatheory as having "a very high level of abstraction" and that it facilitates "thinking about other theories, possibly across disciplines". There has been a renewed call for "good concepts and theories" to stimulate research and counteract a perceived "ambivalent attitude towards theory" in the management fields (Markus & Saunders, 2007). Emerging models of "open innovation" posit that knowledge and resources increasingly reside outside the firm's locus of control. Design and innovation involves putting knowledge into action and has a link to absorptive capacity for new development (the number of graduates and postgraduates involved) (Tsai, 2004). It is more than just acquiring knowledge. A contemporary organisation must become a two-headed Janus looking inward and outward within a complex ecological system, and the teaching of the subject must accommodate this view. Using the influential paper of Thomas Chamberlin as a starting point, a metatheory was developed using abduction as originally proposed by Peirce and more recently presented to the management discipline by Van de Ven. The influential philosopher Franz Brentano insisted that the primacy must be given to description over explanation and promulgated that philosophy develops in several phases. All great periods of philosophical growth are characterised by a predominance of theoretical interest. This chapter was developed in the spirit of Brentano's panoramic insight on the importance of theory to the progress of fields of knowledge (Moran, 2000, p. 33). In Chap. 7, I will propose simulation-action learning (SAL) as a conceptual framework to support the teaching of design and innovation. While I will not offer SAL as a theory per se, the theoretical engagement outlined in this chapter greatly influences its conception.

In his seminal paper on the primacy of resources to an organisation, Wernerfelt (1984) commented that his work was meant "only as a first cut at a huge can of worms" (p. 180). I hope that this chapter can stimulate some debate on the contemporary nature of the theoretical underpinning of design and innovation and its implications for teaching the subject.

Chapter Summary

- Theory is universally regarded as important for scholarship and the understanding of a discipline.
- The question of what is meant by the term *theory* is contentious and widely viewed as difficult to answer.
- *HowStuffWorks* provides answers and explanations of how the world works but theory addresses *WhyStuffWorks*.
- The chapter described what constitutes a theory and what constitutes a good theory.
- Abduction is proposed as a method of theory building particularly suitable for design and innovation.
- The theory of design and innovation was addressed in three ways: first, as a view; second, as an interactive process; and, third, as an ecological system.
- A metatheoretical framework was then presented that incorporates these three perspectives.
- The theoretical insights from this chapter influenced the development of the simulation-action learning (SAL) process which will be presented in Chap. 7.

References

Amabile, T. M., Hadley, C. N., & Kramer, S. J. (2003). Creativity under the gun. In *Harvard business review on the innovative enterprise*. Massachusetts: Harvard Business School Press.

Astley, W. G., & Van de Ven, A. H. (1983). Central perspectives and debates in organization theory. *Administrative Science Quarterly, 28*, 245–273.

Banville, C., & Landry, M. (1989). Can the field of MIS be disciplined? *Communication of the ACM, 32*(1), 48–60.

Barney, J. B. (1991). Firm resources and sustained competitive advantage. *Journal of Management Information Systems, 17*, 99–120.

Benbasat, I., & Zmud, R. (1999). Empirical research in information systems: The practice of relevance. *MIS Quarterly, 23*(1), 3–16.

Bessant, J. (2003). Challenges in innovation management. In L. V. Shavinina (Ed.), *The international handbook on innovation*. Oxford, UK: Elsevier.

Bronfenbrenner, U. (1979). *The ecology of human development: Experiments by nature and design*. Cambridge, MA: Harvard University Press.

Bronfenbrenner, U. (Ed.). (2004). *Making human beings human: Bioecological perspectives on human development*. Thousand Oaks, California: Sage Publications.

Chamberlin, T. C. (1890). The method of multiple working hypotheses. *Science, 15*, 92–96.

Checkland, P., & Holwell, S. (1998). *Information systems and information systems*. Chichester, UK: Wiley.

Chesbrough, H. W. (2003). *Open innovation: the new imperative for creating and profiting from technology* Boston: Harvard Business School.

Coakes, E. (2000). Knowledge management: A sociotechnical perspective. In E. Coakes, D. Willis, & S. Clarke (Eds.), *Knowledge management in the sociotechnical world: The graffiti continues* (pp. 4–14). London: Springer.

Conboy, K. (2009). Agility from first principles: Reconstructing the concept of agility in information systems development. [Ph.D. thesis]. *Information Systems Research, 20*(3), 329–354.

Costello, G. J. (2018). Proposing an innovation-based view of the firm. *Irish Journal of Management, 37*(1), 65–79. https://doi.org/10.2478/ijm-2018-0006.

Costello, G. J. (2010). *Innovation and information systems: A case for ecological systems theory.* PhD Thesis, National University of Ireland, Galway.

Costello, G. J., & Donnellan, B. (2008). Seeking the face of innovation with the ethical compass of Emmanuel Levinas. In G. León, A. Bernardos, J. Casar, K. Kautz, & J. DeGross (Eds.), *IFIP International Federation for Information Processing, volume 287, open IT-based innovation: Moving towards cooperative IT transfer and knowledge diffusion* (pp. 97–117). Boston: Springer.

Cranefield, J., & Yoong, P. (2007). *To whom should information systems research be relevant: The case for an ecological perspective*. Paper presented at the Proceedings of the 15th European Conference on Information Systems (ECIS2007) June 7–9 2007, St. Gallen, Switzerland.

Crowston, K., & Myers, M. D. (2004). Information technology and the transformation of industries: three research perspectives. *The Journal of Strategic Information Systems, 13*(1), 5–28.

Douma, S., & Schreuder, H. (2008). *Economic approaches to organizations* (4th ed.). London, United Kingdom: Prentice Hall.

Dubin, R. (1976). Theory building in applied areas. In M. Dunnette (Ed.), *Handbook of industrial and organisational psychology* (pp. 17–40). Chicago: Rand McNally.

Dubin, R. (1978). *Theory development*. New York: Free Press.

EGFSN. (2017). *"Winning by Design" report by the Expert Group for Future Skills Needs (EGFSN) November 2017.* Available at https://dbei.gov.ie/en/Publications/Publication-files/Winning-by-Design.pdf. Accessed February 2018.

Etzkowitz, H., & Leydesdorff, L. (2000). The dynamics of innovation: From National Systems and "Mode 2" to a Triple Helix of university–industry–government relations. *Research Policy, 29*(2), 109–123.

Fagerberg, J. (2005). Innovation: A guide to the literature. In J. Fagerberg, D. Mowery, & R. R. Nelson (Eds.), *The Oxford handbook of innovation* (pp. 1–26). Oxford, UK: Oxford University Press.

Goffin, K., & Mitchell, R. (2005). *Innovation management: Strategy and implementation using the pentathlon framework*. Houndmills, Basingstoke: Palgrave Macmillan.

Grant, R. M. (1996). Toward a knowledge-based theory of the firm. *Strategic Management Journal, 17*, 109–122.

Gregor, S. (2006). The nature of theory in information systems. *MIS Quarterly, 30*(3), 611–642.

Hookway, C. J. (2005). Peirce, Charles Sanders. In T. Honderich (Ed.), *The Oxford companion to philosophy* (2nd ed., pp. 685–688). Oxford, UK: Oxford University Press.

Hoskisson, R. E., Hitt, M. A., Wan, W. P., & Yiu, D. (1999). Theory and research in strategic management: Swings of a pendulum. *Journal of Management, 25*, 417–456.

Keen, P. (1980). *MIS research: Reference disciplines and a cumulative tradition.* Paper presented at the Proceedings of the First International Conference on Information Systems, Philadelphia.

Keen, P. (1991). Relevance and rigor in information systems research: Improving quality, confidence cohesion and impact. In H. Nissen, H. Klein, & R. Hirschheim (Eds.), *Information systems research: Contemporary approaches & emergent traditions* (pp. 27–49). Amsterdam: North-Holland.

Kraaijenbrink, J., Spender, J.-C., & Groen, A. J. (2010). The resource-based view: A review and assessment of its critiques. *Journal of Management, 36*(1), 349–372.

Lang, S. S. (2005). In Appreciation – Urie Bronfenbrenner. *Association for Psychological Science – Observer*. Available on-line through https://www.psychologicalscience.org/observer/in-appreciation-urie-bronfenbrenner accessed March 2020, 18(11).

Leonard, D. (1998). *Wellsprings of knowledge: Building and sustaining the sources of innovation.* Massachusetts: Harvard Business School Press.

Lundvall, B.-A. (1995). *National Systems of innovation: Towards a theory of innovation and interactive learning*. London: Pinter.

Markus, M. L., & Saunders, C. (2007). Looking for a few good concepts...and theories...for the information systems field. *MIS Quarterly, 31*(1), iii–vi.

McLeroy, K. R., Bibeau, D., Steckler, A., & Glanz, K. (1988). An ecological perspective on health promotion programs. *Health Education & Behavior, 15*(4), 351–377.

Metcalfe, M. (2004). Theory: Seeking a plain English explanation. *Journal of Information Technology Theory and Application, 6*(2), 13–21.

Moran, D. (2000). *Introduction to phenomenology*. London/New York: Routledge.

OECD. (2005). *Oslo manual: Guidelines for collecting and interpreting innovation data* (3rd ed.). Paris: Organisation for Economic Co-Operation and Development, Directorate for Science, Technology and Industry.

Paquette, D., & Ryan, J. (2001). *In Bronfenbrenner's view, a child develops within the system of relationships that encompasses the child's environment. Furthermore, each complex layer influences his or her development.* https://s3.amazonaws.com/academia.edu.documents/44165922/bronfenbrenners_ecological.pdf?response-content-disposition=inline%3B%20filename%3DBronfenbrenners_Ecological_Systems_Theor.pdf&X-Amz-Algorithm=AWS4-HMAC-SHA256&X-Amz-Credential=AKIAIWOWYYGZ2Y53UL3A%2F20200307%2Fus-east-1%2Fs3%2Faws4_request&X-Amz-Date=20200307T173012Z&X-Amz-Expires=3600&X-Amz-SignedHeaders=host&X-Amz-Signature=67134320dd9959b14d3a73729de4dd0a2792efeb19ed1f578b732cfd86cc6726

Pierce, J. L., & Delbecq, A. L. (1977). Organization structure, individual attitudes and innovation. *Academy of Management Review, 2*, 27–37.

Peirce, C. S. (2010). *Writings of Charles S. Peirce: A chronological edition. Vol. 8, 1890–1892/ Edited by the Peirce Edition Project*. Bloomington, IN: Indiana University Press.

Penrose, E. (2009). *The theory of the growth of the firm with a new introduction by Christos N. Pitelis*. Oxford, UK/New York: Oxford University Press.

Pettigrew, A. M. (1987). Introduction: Researching strategic change. In A. M. Pettigrew (Ed.), *The management of strategic change*. Oxford, UK: Basil Blackwell.

Poole, M. S., & Van de Ven, A. H. (2000). Towards a general theory of innovation processes. In A. H. Van de Ven, H. L. Angle, & M. S. Poole (Eds.), *Research on the management of innovation : The Minnesota studies* (Vol. 32, pp. 637–662). Oxford, UK/New York: Oxford University Press.

Rothwell, R. (1994). Towards the fifth-generation innovation process. *International Marketing Review, 11*(1), 7–31.

Ruse, M. (2005). Theory. In T. Honderich (Ed.), *The Oxford companion to philosophy* (2nd ed.). Oxford, UK: Oxford University Press.

Saren, M. (1987). The role of strategy in technological innovation: A re-assessment. In I. L. Mangham (Ed.), *Organizational analysis and development*. Chichester: Wiley.

Sautet, F. (2000). *Entrepreneurial theory of the firm*. London: Routledge.

Schumpeter, J. A. (1934). *The theory of economic development: An inquiry into profits, capital, credit, interest, and the business cycle*. Cambridge, MA: Harvard University Press (1959 printing).

Scruton, R. (2004). *Modern philosophy: An introduction and survey*. London: Pimlico.

Slappendel, C. (1996). Perspectives on innovation in organizations. *Organization Studies, 17*(1), 107–129.

Stein, E. (1989). *On the problem of empathy (The collected works of Edith Stein; vol. 3) translated by Waltraut Stein, Ph.D.* Washington, DC: ICS Publications.

Suppe, F. (1999). Abduction. In R. Audi (Ed.), *The Cambridge dictionary of philosophy* (pp. 651–654). New York: Cambridge University Press.

Sutton, R. I., & Staw, B. M. (1995). What theory is not. *Administrative Science Quarterly, 40*, 371–384.

Svennevig, J. (2001). Abduction as a methodological approach to the study of spoken interaction. *Norskrift, 103*, 1–22.

Teece, D. J., Pisano, G., & Shuen, A. (1997). Dynamic capabilities and strategic management. *Strategic Management Journal, 18*(7), 509–533.

Tidd, J., Bessant, J., & Pavitt, K. (2005). *Managing innovation: Integrating technological, market and organizational change*. Chichester, UK: Wiley.

Tsai, W. (2004). Knowledge transfer in intraorganizational networks: Effects of network position and absorptive capacity on business unit innovation and performance. *Academy of Management Journal, 44*:996–1004. In J. Storey (Ed.), *The management of innovation: Volume II* (pp. 533–562). Cheltenham, UK: Northampton Edward Elgar.

Van de Ven, A. H. (1986). Central problems in the management of innovation. *Management Science, 32*(2), 590–607.

Van de Ven, A. H. (Ed.). (2007). *Engaged scholarship: A guide for organizational and social research*. Oxford, UK/New York: Oxford University Press.

Van de Ven, A. H., Angle, H. L., & Poole, M. S. (Eds.). (2000). *Research on the management of innovation: The Minnesota studies*. Oxford, UK/New York: Oxford University Press.

von Hippel, E. (2005). *Democratizing Innovation*. Massachusetts: The MIT Press.

Wacker, J. G. (1998). A definition of theory: Research guidelines for different theory-building research methods in operations management. *Journal of Operations Management, 16*(4), 361–385.

Walton, R. E. (1987). *Innovating to compete*. San Francisco: Jossey-Bass.

Weick, K. (1989). Theory construction as disciplined imagination. *Academy of Management Review, 14*(2), 516–531.

Wernerfelt, B. (1984). A resource-based view of the firm. *Strategic Management Journal, 5*(2), 99–197.

Whelan, E., Parise, S., Valk, J. D., & Aalbers, R. (2011). Creating employee networks that deliver open innovation. *MIT Sloan Management Review, 53*(1), 37–45.

Whelan, E. (2007). Exploring Knowledge Exchange in Electronic Networks of Practice, *Journal of Information Technology, 22*, 5–12.

Whetten, D. A. (1989). What constitutes a theoretical contribution? *Academy of Management Review, 14*(4), 490–495.

Wolfe, R. A. (1994). Organizational innovation: Review, critique and suggested research directions. *Journal of Management Studies, 31*(3), 405–431.

Yin, R. K. (1994). *Case study research: design and methods* London Sage Publications.

Chapter 5
The Principle of Reflection

5.1 Introduction

Boud, Keogh, and Walker (1985) provide a stimulating primer on the subject of reflection; and in the introduction to their collection of articles, they ask a salient question: What turns experience into learning? They summarise their understanding of reflection as "an active process of exploration and discovery, which often leads to unexpected outcomes" (ibid., p. 7). Consistent with a theme of this book (see Chap. 2), these authors trace the notion of reflection to Aristotle's discussion of practical judgement and moral action. Furthermore, they identify John Dewey as the most influential recent figure in the area and in particular his use of the concept of "reflective activity" in learning. In their work, they pinpoint three aspects that should be borne in mind when considering the process of reflection (ibid., p. 11).

- First, and self-evidently, only learners themselves can learn, and only they can reflect on their own experiences. Teachers can intervene in various ways to assist, but they only have access to individual's thoughts and feelings through what individuals choose to reveal about themselves.
- Second, reflection is pursued with intent. It is not idle meanderings or daydreaming but purposive activity directed towards a goal.
- Third, the reflective process is a complex one in which both feelings and cognition are closely interrelated and interactive.

This chapter proposes that habit of reflection-in-action can be embedded in teaching practice through regular use of agile methodologies and use of reflective questionnaires, which dovetails with the characteristics identified by Boud, Keogh and Walker.

The Oxford dictionary defines reflection as giving something "serious thought or consideration" (ODE, 2006). Pearson and Smith (1985) contend that reflection lies at the heart of experience-based learning and in the absence of reflection, experiences remain experiences, and the full potential for learning may not come to

G. J. Costello, *The Teaching of Design and Innovation*, Contemporary Issues in Technology Education, https://doi.org/10.1007/978-3-030-41380-4_5

fruition. Furthermore, they argue that reflection does not have to be a solitary activity, which is important for a work that advocates a team approach to the teaching of design and innovation. Pearson and Smith propose the notion of debriefing as a planned way for reflection to take place in group settings, and they provide guidelines for effectively accomplishing this process (ibid., p. 83). Walker (1985) encourages the powerful role of writing in order to reflect calmly on experiences, an exercise that has been promoted by luminaries such as Carl Jung. There are many methods of writing that assist reflection such as journals, diaries, record books, portfolios and logs, and his preference is for "portfolio". He points out that historically, reflective writing has had a lasting contribution to culture, in examples such as the *Meditations* of Marcus Aurelius and the *Pensées* of Blaise Pascal. Similarly, I will add Dante's reflective journey in his *Commedia* which is regarded as one of the great contributions to literature and indeed civilisation.

Recently, Cunliffe (2002) has argued convincingly for the need for reflective practice and proposes to "reconstruct learning as reflective/reflexive dialogue" (p. 35). The place of dialogue in reflection is taken up later in this chapter, and a case study of its applicability to practitioner learning is outlined in Chap. 6. There is wide agreement in the literature that *reflection* is critical to meeting the dual mandate of action research (AR): addressing a real-life problem through intervention together with the research objective of making a contribution to knowledge (Avison, Lau, Myers, & Nielsen, 1999; Baskerville & Myers, 2004; Coghlan & Brannick, 2005; Davison, Martinsons, & Kock, 2004). One persistent bone of contention has been the paucity of methodological guidance for conducting and evaluating action learning and action research studies especially on the subject of reflection (Costello, Conboy, & Donnellan, 2015). This resulted in Davison et al. (2004) developing a number of principles and assessment criteria that include the *Principle of Learning through Reflection* to address ongoing concerns with lack of rigour. In this chapter, I will argue that, despite the primacy given to this topic, there is still little practical guidance on how to carry out the rather nebulous process of *reflection*. My aim is to contribute to filling this gap by providing guidance for students, lecturers and practitioners. Furthermore, I will propose an addition to Davison et al.'s "Principle of Learning through Reflection" through modifying a set of questions which were based on engagement with the psychology literature (Dick, 2019a).

The chapter is structured as follows. Firstly, a literature review of action learning and action research is presented that examines the role of *reflection* in the approach. I will then argue for the importance of dialogue and listening (Knights, 1985) for reflection and follow this with an overview of Donald Schön's seminal work on *reflection-in-action*. The next section proposes that the phenomenology of Edmund Husserl provides a philosophical underpinning for action learning and discusses the theme of reflection in the phenomenological literature. Following this, the chapter provides a novel pedagogical suggestion on embedding the habit of reflection-in-action in teaching practice through regular use of agile development methodologies. Finally, a reflective questionnaire is developed to practically assist individuals and teams in carrying out reflection within design and innovation projects.

5.2 Action Learning and Action Research

In this section I will provide a brief overview of action learning and its associated methodology – action research – and I will emphasise the importance of reflection drawing on the influential work of Donald Schön's reflection-in-action.

5.2.1 Action Learning

Experiential learning theory can be traced to the seminal work by Kolb and Fry (1975) in MIT who advocated the principle that the learner is a creator of learning rather than the passive recipient of information. According to Pedler and Abbott (2013), action learning per se originated with Reginald Revans (2011), and these authors propose four elements of action learning that are presented in Fig. 5.1. Furthermore, Weinstein (1999) concludes that reflection is a key component to the action learning cycle, while Marquardt (2004) emphasises the use of a coach in action learning. This latter point will be taken up in the discussion of the role of the teacher in the practices of design and innovation teaching in Chaps. 6 and 7.

McGill and Brockbank (2004) define action learning as follows:

> Action learning is a continuous process of learning and reflection that happens with the support of a group or 'set' of colleagues, working on real issues, with the intention of getting things done. The voluntary participants in the group or 'set' learn with and from each other and take forward an important issue with the support of the other members of the set.

Their definition dovetails with the aim of this book: to provide guidance for the development of student teams who work on real-world design briefs and in the process learn, with and from each other, and where the teacher chiefly takes the role of a coach and mentor.

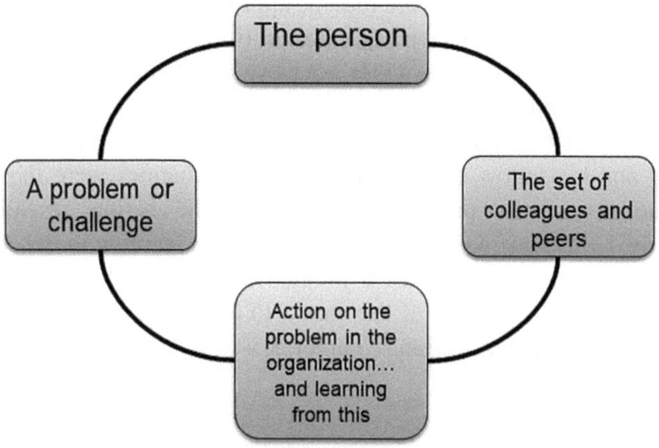

Fig. 5.1 Four elements of action learning. (From Pedler, M. and Abbott, C. (2013, p. 10))

Revans (2011) acknowledges his debt to Aristotle by stating that managers as action-reflectors and learner-doers become disciples of Aristotle by "doing what they set out to do, and by setting out to do what they believe to be worth doing". He goes on to say:

> Aristotle, too, has bequeathed to us a system of ethics balanced upon action and reflection; one gets to understand by doing, and to do by understanding.

Brook, Pedler and Burgoyne (2012, p. 271) in their examination of Revans' work identify some key principles that conventional (or classical) *action learning* adheres to:

- The requirement for action as a basis for learning
- Profound personal development resulting from reflection upon action
- Working with problems (no right answers) not puzzles (susceptible to expert knowledge)
- Problems being sponsored and aimed at organisational as well as personal development
- Action learners working in sets of peers ("comrades in adversity") to support and challenge each other
- The search for fresh questions and "Q" (questioning insight) takes primacy over access to expert knowledge or "P"

Kelliher (2014) examined the experience and outcomes of applying action learning in an executive MBA programme. Her *Change Management* module encompassed the principles of critical thinking and reflective practice, with a discrete action learning approach to the delivery of the curriculum and completion of assignments. According to Kelliher, the action learning philosophy facilitates skill development by integrating knowledge gained from experience and knowledge gained from theory and the literature. A focus of the module is critical reflection among the MBA students engaged in the action learning cycle. Furthermore, the learning was applied through a year-long project applied in the practitioner's work environment. The research indicated that the approach of completing a change management intervention in their own organisation (the action) supported by a project supervisor working as part of a wider advisory panel (the learning) had positive benefits for the students and for their workplaces.

Kolb and Fry (1975) propose that arguably the most seminal contributions of Kurt Lewin and his fellow researchers on group dynamics were the experiential learning model and its practical counterpart, the action research method. Consequently, I will now place the approach of action learning in the context of the development of action research as a method and a discipline. In addition, this addresses one of the objectives of this book: to contribute a pedagogical research agenda for design and innovation. Chapter 8 will discuss the formation of such an agenda.

5.2.2 Action Research

Action research (AR) originated from the work of Kurt Lewin during the 1940s and has been summarised as an approach that "combines theory and practice (and researchers and practitioners) through change and reflection in an immediate problematic situation within a mutually acceptable ethical framework" (Avison et al., 1999). The application of AR has not been without controversy particularly in debates with positivist science on the justification and generation of knowledge. These arguments were addressed by Susman and Evered (1978) in their influential description of AR as consisting of a cyclical process involving five phases: diagnosing, action planning, action taking, evaluating and specifying learning. The focus of AR is to address real-life problems through intervention together with the research objective of making a contribution to knowledge. In the realm of management information systems research, Avison et al. (1999) argued that it took until 1998 for the community to agree that qualitative approaches, such as action research (AR), were finally gaining acceptance and proposed that "to make academic research relevant, researchers should try out their theories with practitioners in real situations and real organizations". Coghlan and Brannick (2005, p. 125) emphasise the importance of the social and academic context in which action research is carried out, and as a result, the contextual setting of a practitioner case study is presented later. This theme is echoed in the work of Bob Dick (2019b) which will have significant influence on the arguments of this chapter. Dick, an academic working in the field of psychology, proposes that the AR methodology has the twofold aim of action and research:

- *Action* designed to bring about change in some community, organisation or programme
- *Research* to increase understanding on the part of the researcher or the client or both – and in many cases some wider community

Reason and Bradbury aim to "draw together some of the main threads that form the diverse practices of action research" and propose an almost lofty vision of AR contributing to the world's well-being and sustainability, in areas ranging from the economic and political to the psychological and spiritual. The following quotation with its emphasis on understanding and reflection is of particular relevance to this chapter (Reason & Bradbury, 2001, p. 2).

> So action research is about working towards practical outcomes, and also about creating new forms of understanding, since action without reflection and understanding is blind, just as theory without action is meaningless.

5.3 Importance of Reflection in AR

The process of reflection is integral to AR and is emphasised in the literature (Avison et al., 1999; Baskerville & Myers, 2004; Coghlan & Brannick, 2005; Davison et al., 2004). Braa and Vidgen (2000) make the salient point that in the course of research,

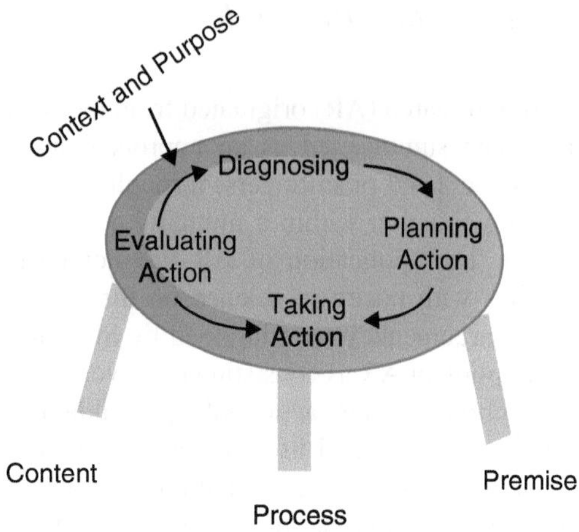

Fig. 5.2 Meta-cycle of inquiry. (From Coghlan and Brannick (2005))

in addition to learning from the research content, there should also be learning about the process of inquiry. In relation to this, Coghlan and Brannick (2005, p. 25), drawing from a number of antecedent publications by authors such as Argyris and Mezirow, propose that this "reflection on reflection" results in "learning about learning". They call this process *meta-learning* which consists of three types of critical reflection:

- *Content reflection*: this is where you think about the issues and what is happening
- *Process reflection*: this is where you think about strategies and procedures and how things are being done
- *Premise reflection*: this is where you critique underlying assumptions and perspectives

Coghlan and Brannick then superimpose these three constructs on their version of the action research cycle to develop a meta-cycle of inquiry, which is shown in Fig. 5.2. Furthermore, they contend that the dynamic of "reflection on reflection" differentiates action research from straightforward problem-solving. They also make the point that focusing on a *problem* can lead to convergent thinking while focusing on an *opportunity* can lead to divergent thinking. Overemphasis on either can be to the detriment of a design and innovation project where creativity and planning need to be effectively orchestrated. The antidote, they propose, is *issue* identification and selection (Coghlan & Brannick, 2014, p. 66).

Coghlan and Brannick build on Argyris and Mezirow as follows (2005, p. 24):

- The content of what is diagnosed, planned, acted on and evaluated is studied.
- The process of how diagnosis is undertaken, how action planning flows from that diagnosis and is conducted, how closely the implemented actions follow the stated plans and how evaluation is conducted are critical foci for inquiry.
- The premise reflection consists of an inquiry into the unstated, and often nonconscious, underlying assumptions which govern attitudes and behaviour. For example,

the organisation's culture can have a profound impact on viewing and discussing issues without team members being aware of them.

Now that I have presented my argument on the importance of reflection to action learning and action research, I will now outline some work on the importance of dialogue to the reflective process. Furthermore, the work will be used in the analysis of the practitioner case study in Chap. 6.

5.4 Dialogical Action Research

Mårtensson and Lee (2004, p. 507) have suggested and described a novel form of action research called *dialogical* AR wherein "the scientific researcher does not 'speak science' or otherwise attempt to teach scientific theory to the real-world practitioner, but instead attempts to speak the language of the practitioner and accepts him as the expert on his organisation and its problems". In their paper Mårtensson and Lee (2004, p. 531) propose that "reflective one-to-one dialogues" between the practitioner and the researcher, which take place at regular intervals in a location removed from the organisation, can help the manager to "reflect on, learn from, and remedy managerial problems in the organisation". In their schema, the role of the researcher consists in suggesting actions based on one or more theories taken from their discipline. The implementation of these suggestions is left to the judgement of the practitioner based on their experience, expertise and tacit knowledge together with their reading of the organisational situation that confronts them. Furthermore, the ongoing dialogue is presented as an *interface* between the scientific world of the researcher, marked by *theoria*, and the everyday world of the practitioner, which is marked by *praxis*. The overall aim of dialogical AR is to bring about some improvement to the real-world problem of the practitioner while at the same time contributing to the development, confirmation or disconfirmation of theory by the researcher. Mårtensson and Lee (2004) draw heavily on Schön's (1990) model of professional inquiry consisting of a pattern of five features: situation requiring attention, a surprising response, reflection-in-action, critical examination and restructuring and an "on-the-spot experiment" (Mårtensson & Lee, 2004, p. 510). They make a fundamental distinction between traditional forms of consulting and dialogical AR in that the latter always involves reflection and learning. Furthermore, unlike consulting, AR involves someone who has academic expertise rooted in some scientific discipline, where teamwork takes place between researcher and practitioner and where "negative feedback" is seriously taken on board.

In line with the suggestion by Coghlan and Brannick (2005), quoted above, Mårtensson and Lee (2004, p. 514) emphasise that it is incumbent on the researcher to "explicitly and intentionally acquire an understanding of the social and historical context of the organisation and its problems". As a result, the design and innovation context is an important element of the case studies outlined in Chap. 6. In their vision of dialogical AR, the scientist makes suggestions to the practitioner, but the practitioner remains the "agent of action" using their explicit and tacit knowledge

(Mårtensson & Lee, 2004, p. 515). Furthermore, Mårtensson and Lee (2004) see the role of the researcher as having the following attributes in the one-on-one dialogues: first, to listen in order to identify the problem that requires some action; second, to gather the facts to form the basis of deciding what suitable theory can be applied to the problem area; and, third, to suggest and monitor appropriate actions to the practitioner. Interestingly, Mårtensson and Lee (2004) use the analogy of an anthropologist spending time to understand the world of the natives, i.e. the practitioner. A case study of the application of dialogical AR carried out during the implementation of an innovation project in a multi-national corporate (MNC) subsidiary will be presented in Chap. 6. Now I will introduce the work of one of the most influential contributors to the modern concept of reflection in practice.

5.5 Schön's Reflection-in-Action

Donald Schön's (1983) publication of *The Reflective Practitioner* is regarded as a seminal work in the debate on the benefits of *reflection* for practice and research. In the book he criticises the prevailing academic epistemology as having nothing to offer either practitioners "who wish to gain a better understanding of the practical uses and limits of research-based knowledge" or scholars "who wish to take a new view of professional action". Schön begins with the assumption that "competent practitioners usually know more than they can say" and that they exhibit "a kind of knowing in practice, most of which is tacit". Furthermore, in disciplines such as medicine, management and engineering, his experience was that professionals were exhibiting "a new awareness of a complexity which resists the skills and techniques of traditional expertise". Schön presents the academic about-turn of Russell Ackoff, one of the founders of the discipline of operations research, as powerful evidence of this shift. Ackoff had recently described the dynamic and complex situations faced by managers as being akin to *messes*, which did not lend themselves to the problem-solving techniques of mathematical models and algorithms. Schön argues that this dominant epistemology of practice is based on the model of *technical rationality* where "professional activity consists in instrumental problem-solving made rigorous by the application of scientific theory and techniques". Its origins lie in the rise of the technological programme that came to dominate Western society in the nineteenth century. This resulted in Auguste Comte formulating his philosophy of *positivism* which contains three principal doctrines (Schön 1983, p. 32):

- Empirical science is not just a form of knowledge but the only source of positive knowledge of the world
- Men's minds need to be cleansed of mysticism, superstition and other forms of pseudo-knowledge
- Extending of scientific knowledge and technical control to human society in order to make technology "no longer exclusively geometrical, mechanical or chemical, but also primarily political and moral".

Schön then laments that the seeds of positivism were firmly planted in the curricula of American universities and professional schools, a factor, which he argues, has contributed significantly to the contemporary fissure between research and practice. Furthermore, he concludes that the present difficulty in accommodating contemporary phenomena such as "complexity, uncertainty, instability, uniqueness, and value conflict" stems from the positivist origins of technical rationality. He proposes the primacy of *problem-setting* over *problem-solving* for practitioners. Problem-setting he defines as an interactive process in which "we *name* the things to which we will attend and *frame* the context in which we will attend to them". This theme is taken up in Chap. 6 where students are immersed within a design scenario or setting rather than just presented with a defined "problem" to be solved.

The perennial dilemma of *rigour and relevance* is presented using the analogy of a hilly landscape. He describes the "high hard ground" as the place where practitioners can effectively apply research-based theories and methods. However, the important and challenging problems exist in the "swampy lowland" of messy situations that do not respond to neat technical solutions. Furthermore, according to Schön the earlier models of *technical rationality* have in general "failed to yield effective results" when dealing with the complex and fuzzy problems of business management.

In order to fit practice into the models of technical rationality and deal with the tension of rigour versus relevance, practitioners become "selectively inattentive" to data that do not fit neatly into their predefined categories. For example, he states that "designers of management information systems" frequently fail to notice that in reality "their systems trigger games of control and evasion". In addition, the following comment by Schön seems pertinent to the philosophical debate within the management disciplines: "among philosophers of science no one wants any longer to be called a Positivist". Furthermore, he observes that the growing rebirth of many areas recently consigned to the positivist graveyard such as craft, artistry and myth is further evidence of the failure of the positivist programme. However, he is at pains to point out that his problem is not with science per se but on the view of science portrayed by positivism. The influence of positivism on contemporary scholarship will be discussed in Chap. 8.

As an antidote to *technical rationality*, Schön proposes *reflection-in-action* built on the idea of *knowing-in-action*, which he explains as:

> Our knowing is ordinarily tacit, implicit in our patterns of action and in our feel for the stuff with which we are dealing. It seems right to say that our knowing is *in* our action.

Furthermore, the "common sense" that reveals *knowing-in-action* to us also reveals that sometimes we "think about what we are doing". Characteristics of this reflection-in-action include many colloquialisms such as:

- Thinking on your feet
- Keeping your wits about you
- Learning by doing
- And what baseball pitchers would call *finding the groove*

The art of reflection-in-action results in a practitioner – when faced with a challenging situation – setting the problem in a new context or frame which he calls a *frame experiment*. The overall design of the student experience outlined in Chap. 6 and the proposal of simulation-action learning (SAL) in Chap. 7 are framed to prepare students to be innovative and enterprising in challenging modern workplace environments.

One conclusion of Schön that I find contentious is his proposition that when a practitioner reflects-in-action, he does not depend "on the categories of established theory and techniques but constructs a new theory of the unique case". This would seem to contradict the legitimate viewpoint of many philosophers such as Quine, who contend that at the most basic level, our language and sentences are based on numerous underlying theories. Schön believes that reflection-in-action is still not generally accepted in professional practice, even by those who actually carry it out, due to the professions still being viewed solely in terms of their technical expertise. He begins to describe an epistemology of reflection-in-action that "accounts for artistry in situations of uniqueness and uncertainty" to deal with conditions where the model of technical rationality "appears as radically incomplete". A concern that I have with his initial work is that while he provides a convincing *deconstruction* of positivism, he does not justify any philosophical alternative to underpin reflection-in-action. However, in a subsequent publication on *Educating the Reflective Practitioner* (1990), Schön acknowledges his debt to John Dewey and to a number of others including Kurt Lewin, which is a relevant connection with the theme of action learning and research. Concepts used by Schön in *Educating the Reflective Practitioner* will be examined in Chap. 6, where certain routines were established in the case studies with the aim of instilling the habit of reflection-in-action in a student cohort. Furthermore, Schön's concept of the *practicum* is utilised in the development of the simulation-action learning (SAL) theoretical framework. Now I will provide an overview of a methodology that I propose to use in order to implement reflection-in-action in the practice chapters of the book.

5.5.1 Agile Design and Innovation

Agile development has gained much prominence in the software industry in recent years (agilemanifesto, 2001; Conboy, 2009). In a wide-ranging view of the literature, Abrahamsson, Salo, Ronkainen, and Warsta (2002) caution that while there has been growing interest in the concept from both industry and academia, there is still a lack of agreement on what "agile" actually means. A core agile development methodology is called Scrum. It is interesting to note that Abrahamsson et al. trace the first use of the Scrum concept to literature on product development from the mid-1980s in Japan. Scrum involves cross-functional self-organising teams of five to ten people. Each development stage typically consists of several short development cycles, known as sprints. Cervone (2011, p. 21) describes the Scrum sprints as follows: in many projects, but not all, each sprint begins with a daily Scrum meeting. This meeting, typically lasting no more than 15 minutes, is held every day between the Scrum master (who chairs the meeting) and the Scrum team. In this meeting, every team member briefly answers three questions:

1. What did you do since the last Scrum?
2. What are you doing until the next Scrum?
3. What is stopping you getting on with your work?

Recently, Robert Cooper, a seminal contributor to the stage-gate product development process, has described how firms have integrated elements of the agile IT product development method into their traditional gating processes to develop physical products (Cooper, 2016). Sommer, Hedegaard, Dukovska-Popovska, and Steger-Jensen (2015) conducted a study of seven technology-intensive companies and concluded that integrating agile models with the stage-gate process can help improve new product development performance. Furthermore, an output from their research was the proposition of an agile/stage-gate hybrid called the "Industrial Scrum" framework.

The cases presented in the next chapter will describe the use of the Scrum methodology with student teams using the three Scrum questions outlined above. The discipline of regular Scrum meetings contributes to building the habit (à la Aristotle) of reflection-in-action among the students. Furthermore, I have experimented with dividing the Scrum meetings into two interrelated sections: the *project Scrum* and the *reflection Scrum*.

I will examine a philosophical foundation for action learning in the next section. It is interesting to note that the seminal action research authors Susman and Evered (1978, p. 596) propose phenomenology as a philosophical viewpoint that legitimates action research.

5.6 Philosophical Underpinning

In this chapter I have recognised the influence of Kurt Lewin, widely regarded as the originator of action research, on active teaching approaches and disciplines such as management research. Urie Bronfenbrenner (1979), who regarded Lewin as an important mentor for his work, firmly places him in the tradition of phenomenology. Consequently, I will now seek to locate the philosophical underpinnings of the chapter in the phenomenological movement initiated by Edmund Husserl. This continues an objective of the book, begun in Chap. 2, to support the teaching of innovation with philosophical principles.

5.6.1 The Phenomenology of Edmund Husserl

Edmund Husserl, the founding father of phenomenology, is regarded as having instigated one of the most important philosophical movements of the twentieth century (Grossmann, 2005). The system has had an immense influence in Europe in areas spanning psychology, law, values, aesthetics and religion (Inwood, 2005). Husserl's original studies were in the area of mathematics, and his most influential teacher was the philosopher Franz Brentano. His philosophy underwent a transition from his earlier studies on the "phenomenology of mathematical and logical concepts" to the "transcendental idealism" developed in his later major work *Ideas: General Introduction to Pure Phenomenology* (Elveton, 1970). Lauer (1965) argues that with the passage of time, a precise definition of "phenomenology" became more difficult but proposed that the term could be traced back to a "distinction made by Kant between phenomenon or appearance of reality in consciousness, and the noumenon, or being of reality itself". However, he points out that Husserl rejected what he perceived as the "dualism" of Kant. Lauer continues to explain the phenomenology of Husserl as both a method and a philosophy. Method in so far as it provides the steps that must be followed "to arrive at the pure phenomenon, wherein is revealed the very essence not only of appearances but also of that which appears". In the realm of philosophy, "it claims to give necessary, essential knowledge of that which is". Thus, phenomenology advocates a "return to things because a thing is the direct object of consciousness in its purified form". This approach was in opposition to "illusions, verbalisms or mental constructions" implied by many contemporary movements. Moran (2000, p. 15) argues that the major contribution of phenomenology to contemporary philosophy is its conception of "objectivity-for-subjectivity" and one of the aspects of the early work of Husserl was its grounding in Realism. Several of his pupils and collaborators in Göttingen, such as Max Scheler, Edith Stein and Roman Ingarden, were somewhat disappointed by Husserl's "turning" towards idealism in his later work and continued to identify with the realism of the early Husserl.

5.6.2 The Place of Reflection in the Phenomenological Approach

Husserl considered that philosophy should be carried out as a rigorous science using the structured methodology of reason and his vision was that the phenomenological approach (of bracketing the natural world and a reduction to pure consciousness) could overcome and synthesise the radical disagreements of contemporary philosophy. Following Brentano, he held the conviction that the fundamental purpose of philosophy is in *description* and not causal explanation. The first edition of Husserl's book *Logical Investigations* published in 1900 catapulted him into the top echelon of German philosophy. Moran points out that while the first edition equated phenomenology with descriptive psychology, Husserl began to distance his evolving philosophy from any type of psychology. The objective of phenomenology was to focus exclusively on the meaning-constituting function of acts. Furthermore, Moran (2000, p. 77) describes the central importance of reflection as follows:

> Phenomenology proceeds by a pure 'intuiting' (anshauen) and 'reflection' (Reflexion) which "precludes any co-positing of objects alien to consciousness.

In this early work, Husserl proposed that the way to get at the "pure features of consciousness" is called reflection (*Reflexion*). However, in his later influential book *Ideas I* while still retaining the idea of reflection, he formulated the more technical concept of reduction. A more detailed discussion of reflection in the work of the early phenomenologists is outside the scope of this chapter. Nevertheless, I suggest that this section has located the central notion of *reflection*, in a philosophical tradition that is closely linked with action research and action learning.

5.7 Criteria for Learning Through Reflection

I will now address my critique of the literature on reflection, namely, that there is a paucity of guidance on how to carry out the process in real-world or simulated situations – either with students or practitioners. To address this situation, I will propose guiding checklists that can be applied to stages of an action learning module. These can be used by teachers and students (or in the case of continuing professional development – by practitioners) to evaluate and reflect on the case undertaken.

The first checklist takes one of the five evaluation principles proposed by Davison et al. (2004): the Principle of Learning through Reflection. This principle is considered by Davison et al., following Lau (1997), as the "most critical activity in AR". The criteria adapted from these authors to the teaching of design and innovation is outlined in Table 5.1.

Now I will propose a module questionnaire that can be used by students and teachers to distil learning through reflection at the end of a project. This exercise could be categorised essentially as reflection-on-action. Consequently, in Chap. 6 I will outline how I endeavour to instil the ongoing discipline of reflection-in-action through regular use of agile development methodologies such as Scrum (Scrum, 2019). However, they should not be regarded as distinct approaches since the overall strategy is develop an ethos of reflection in the learners (Table 5.2).

Table 5.1 Criteria for the Principle of Learning through Reflection

Did the students provide progress reports to the teacher and other stakeholders – such as clients involved in a simulated real-world design and innovation project?
Did both the students and the client reflect upon the outcomes of the project?
Were the student activities and outcomes reported clearly and completely?
Were the results considered in terms of implications for further action in this situation, e.g. real-world implementation of the design/innovation or follow-on projects?
Were the results considered in terms of implications for action to be taken in related learning domains, e.g. the future delivery of the module?
Were the results considered in terms of implications for the teaching and learning community (general knowledge, informing/re-informing theory)?
Were the results considered by the teacher in terms of the general applicability to student or practitioner learning?

Adapted from Davison et al. (2004)

Table 5.2 End of module questionnaire to assist reflection

Evaluation of project actions and learning outcomes Purpose of this review is to answer the questions: ***What was the action? What were the outcomes (results)? What was learned?***		
Questions adapted from (Dick, 2019a) *and* (Davison et al., 2004).		
Section A. The action (s)		
A1	What was the main action(s) taken during the project?	Intended Unintended
Any other reflections that you think appropriate to include here?		
Section B. The outcomes (results)		
B1	Did I get the outcomes I wanted? (or more realistically, what were the outcomes I got and how well do these accord with those I sought?)	Intended Unintended
B2	To the extent that I got them, do I still want them (i.e. are they any use)? Why or why not?	
B3	To the extent that I didn't get them, why not?	
B4	Did I succeed in carrying out the planned actions? If not, what prevented our discouraged me?	
Any other reflections that you think appropriate to include here?		
Section C. Learning		
C1	When faced with a similar situation in future: Would I try to pursue different outcomes based on what I have learned in this project?	
C2	Would I try different actions to pursue similar outcomes?	
C3	What have I learned about myself, my skills, my attitudes, etc.?	
C4	What are the changes to my perceptions and knowledge about the topic of management?	
Any other reflections that you think appropriate to include here?		
Section D. Future work		
D1	What are the implications for further actions in this situation? For example, do I have any practical suggestions for further changes or follow-on projects in the organisation? Do I have any suggestions for changes or for new structures or systems?	
D2	What is my feedback (positive and negative) for the research process?	
D3	Do I see value in future similar research collaboration and have I any suggestions?	
Any other reflections that you think appropriate to include here?		

5.8 Conclusions

Kolb (1984) traced the intellectual ancestry of experiential learning to Dewey, Lewin and Piaget and proposed the approach as a way of revitalising third-level curricula and of coping with the many challenges facing contemporary higher education. However, he cautions throwing out the baby with the bathwater and sees the future of education being realised in "a spirit of cooperative innovation that integrates the best of the traditional and experiential" (ibid., p. 5). Experiential learning is, according to Kolb, the process that links education, work and personal development. The translation of experience into learning is generally facilitated by the process of reflection. The central objective of this chapter was to address the paucity of guidance for teachers and students of design and innovation on how to carry out the process of *reflection*. This task is integral to the methodology of experiential and action learning. The theoretical basis of the chapter was built on the work of David Kolb and on Donald Schön's reflection-in-action with its philosophical underpinning in the work of Edmund Husserl. A structured questionnaire was developed that is designed to stimulate the reflective process and assist in the metamorphosis from merely engaging in a hands-on project to enabling cognitive learning from such an engagement. In addition, the chapter proposed that implementation of agile development methodologies such as Scrum facilitates the teaching of reflection-in-action. The case studies in Chap. 6 will utilise these guides and provide empirical examples of instilling the practice of reflection-in-action in an educational setting using "agile" approaches.

Chapter Summary

- Reflection is critical to converting experience into learning.
- There is a paucity of guidance on how to engage in reflective activity and to imbed reflection in the learning process.
- Experiential learning theory has been influenced by the work by Kolb and Fry who advocated the principle that the learner is a creator of learning rather than the passive recipient of information.
- Another major influence has been Kurt Lewin, regarded as a father of action learning and action research, who set out a path for combining both action and learning.
- Donald Schön's publication of *The Reflective Practitioner* is regarded as a seminal work in the debate on the benefits of reflection for practice and research.
- Reflective dialogues have demonstrated benefits for the process of reflection.
- A checklist and questionnaire were developed from the literature to provide teachers and students of design and innovation with a structured approach to inculcating reflection in the learning process.
- The empirical work in Chap. 6 will outline how agile development methodologies such as Scrum complement this questionnaire in order to directly address the challenge of embedding reflection-in-action in the curriculum.

References

Abrahamsson, P., Salo, O., Ronkainen, J., & Warsta, J. (2002). *Agile software development methods: Review and analysis*. Aailable on-line through https://www.vtt.fi/inf/pdf/publications/2002/P478.pdf. Accessed October 2019. Espoo, Finland: VTT publication 478.

agilemanifesto. (2001). *Manifesto for Agile Software Development*. Available on line through http://agilemanifesto.org/. Accessed July 2019.

Avison, D. E., Lau, F., Myers, M. D., & Nielsen, P. A. (1999). Action research. *Communications of the ACM, 42*(1), 94–97.

Baskerville, R., & Myers, M. D. (2004). Special issue on action research in information systems: Making IS research relevant to practice—Foreword. *MIS Quarterly, 28*(3), 329–335.

Boud, D., Keogh, R., & Walker, D. (Eds.). (1985). *Introduction: What is reflection in learning*. London: RoutledgeFalmer.

Braa, K., & Vidgen, R. (2000). Research: From observation to intervention. In K. Braa, C. Sørensen, & B. Dahlbom (Eds.), *Chapter 12. Planet internet*. Studentlitteratur, Lund, Sweden. Previously published as: Braa & Vidgen, (1999) Interpretation, intervention, and reduction in the organizational laboratory: A framework for in-context information system research. *Accounting, Management and Information Technologies,* (9): 25–47. Elsevier Science Ltd.

Bronfenbrenner, U. (1979). *The ecology of human development: Experiments by nature and design*. Cambridge, MA: Harvard University Press.

Brook, C., Pedler, M., & Burgoyne, J. (2012). Some debates and challenges in the literature on action learning: The state of the art since. *Revans Human Resource Development International, 15*(3), 269–282. https://doi.org/10.1080/13678868.13672012.13687926.

Cervone, H. F. (2011). Understanding agile project management methods using Scrum. *OCLC Systems & Services: International Digital Library Perspectives, 27*(1), 18–22. https://doi.org/10.1108/10650751111106528.

Coghlan, D., & Brannick, T. (2005). *Doing action research in your own organization* (2nd ed.). London: Sage Publications.

Coghlan, D., & Brannick, T. (2014). *Doing action research in your own organization* (4th ed.). London: Sage Publications.

Conboy, K. (2009). Agility from first principles: Reconstructing the concept of agility in information systems development. [Ph.D. thesis]. *Information Systems Research, 20*(3), 329–354.

Cooper, R. G. (2016). Agile–Stage-Gate Hybrids. *Research-Technology Management, 59*(1), 21–29.

Costello, G. J., Conboy, K., & Donnellan, B. (2015). *Reflections on "reflection:" In action research*. Paper presented at the Uncertainty is a Great Opportunity.15th European Academy of Management Conference (EURAM) 17–20 June 2015,Warsaw.

Cunliffe, A. L. (2002). Reflexive dialogical practice in management learning. *Management Learning, 33*(1), 35–61.

Davison, R. M., Martinsons, M. G., & Kock, N. (2004). Principles of canonical action research. *Information Systems Journal, 14*(1), 43–63.

Dick, B. (2019a). *Questions for critical reflection*. Available at http://www.aral.com.au/. Accessed December 2019.

Dick, B. (2019b). *You want to do an action research thesis? You want to do an action research thesis?* Available at http://www.aral.com.au/. Accessed December 2019.

Elveton, R. O. (1970). Introduction. In R. O. Elveton (Ed.), *The phenomenology of Husserl: Selected critical readings/edited, translated, and with an intro. by R. O. Elveton*. Chicago: Quadrangle Books.

Grossmann, R. (2005). Phenomenology. In T. Honderich (Ed.), *The Oxford companion to philosophy* (2nd ed., pp. 695–697). Oxford, UK: Oxford University Press.

Inwood, M. J. (2005). Husserl, Edmund. In T. Honderich (Ed.), *The Oxford companion to philosophy* (2nd ed., pp. 408–410). Oxford, UK: Oxford University Press.

Kelliher, F. (2014). Just do it: Action learning as a catalyst for reflective practice on an MBA programme. *Procedia - Social and Behavioral Sciences, 141*, 1275–1280. https://doi.org/10.1016/j.sbspro.2014.05.220.

Knights, S. (Ed.). (1985). *Reflection and learning: The importance of a listener*. London: RoutledgeFalmer.

Kolb, D. A. (1984). *Experiential learning : Experience as the source of learning and development*. London Prentice-Hall: Englewood Cliffs.

Kolb, D. A., & Fry, R. E. (1975). Toward an applied theory of experiential learning. In C. Cooper (Ed.), *Theories of group processes*. New York: Wiley.

Lauer, Q. (1965). *Phenomenology : Its genesis and prospect*. New York: Harper & Row.
Lau, F. (1997). A review of the use of action research in information systems studies. In A. Lee, J. Liebenau & J. DeGross (Eds.), *Information Systems and Qualitative Research* (pp. 31–68). London: Chapman & Hall.
Mårtensson, P., & Lee, A. S. (2004). Dialogical Action Research at Omega Corporation. *MIS Quarterly, 28*(3), 507–536.
Marquardt, M. J. (2004). *Optimising the power of action learning*. Palo Alto, CA: Davis-Black.
McGill, I., & Brockbank, A. (2004). *The action learning handbook: Powerful techniques for education, professional development and training*. London: Routledge.
Moran, D. (2000). *Introduction to phenomenology*. London/New York: Routledge.
ODE. (2006). Reflection. In *Oxford Dictionary of English* (2nd ed., Revised). Oxford, UK: Oxford University Press.
Pearson, M., & Smith, D. (Eds.). (1985). *Debriefing in experience-based learning*. London: RoutledgeFalmer.
Pedler, M., & Abbott, C. (2013). *Facilitating action learning: A practitioner's guide*. Maidenhead, UK: McGraw-Hill Education.
Reason, P., & Bradbury, H. (2001). Introduction: Inquiry and participation in search of a world worthy of human aspiration. In P. Reason & H. Bradbury (Eds.), *Handbook of action research: Participative inquiry and practice*. London: Thousand Oaks.
Revans, R. W. (2011). *ABC of action learning*. Farnham, UK: Routledge.
Schön, D. A. (1983). *The reflective practitioner: How professionals think in action*. New York: Basic Books.
Schön, D. A. (1990). *Educating the reflective practitioner: Toward a new design for teaching and learning in the professions*. San Francisco: Wiley.
Scrum. (2019). *Scrum alliance: Transforming the world of work*. Available on-line through https://www.scrumalliance.org/. Accessed October 2019.
Sommer, A. F., Hedegaard, C., Dukovska-Popovska, I., & Steger-Jensen, K. (2015). Improved product development performance through agile/stage-gate hybrids: The next-generation stage-gate process? *Research-Technology Management, 58*(1), 34–44.
Susman, G. I., & Evered, R. D. (1978). An assessment of the scientific merits of action research. *Administrative Science Quarterly, 23*(4), 582–603.
Walker, D. (Ed.). (1985). *Writing and reflection*. London: RoutledgeFalmer.
Weinstein, K. (1999). *Action learning: A practical guide*. Aldershot, Hampshire, UK: Gower.

Part II
Practices

Chapter 6
Cases of Teaching Design and Innovation

6.1 Introduction

This chapter will describe two undergraduate case studies that aimed to instil design and innovation competences in students of both engineering and business disciplines. The cases involved student teams working with multi-national subsidiaries and incubation centre start-ups. One of the objectives of this book is to assist continuing professional development (CPD) in the area of design and innovation. Consequently, a third case is presented that involved the author's collaboration with a practitioner who aimed to transform his organisation from being a traditional manufacturing location into an innovative team-based location. This case is presented in the context of research-informed teaching.

Initially the chapter will provide an example of using student-entrepreneur collaboration in the teaching of modules in the product design stream of mechanical engineering students in an Institute of Technology, located in the West of Ireland (Costello, 2014). Following this, case study 2, will describe how cross-functional engineering and business teams collaborated in design and innovation projects entered into a national student enterprise competition (Costello, 2016).

According to Boud and Feletti (1998, p. 1), "problem-based learning is the most significant innovation in the area of education for the professions in many years". The focus in this type of learning is to provide the students with problem scenarios so that they can learn through a process of action and reflection (Savin-Baden, 2003). However some scholars argue that design "is hard to learn and harder still to teach"(Dym, Agogino, Eris, Frey, & Leifer, 2005). Furthermore organisations, such as Engineers Ireland, are calling for graduate engineers to have more rounded skills in the areas of presentation, communication and team-working (Engineers Ireland, 2019). The application of problem- or project-based learning to the engineering discipline is frequently traced back to the Danish tradition established in two universities in the 1970s. One of these, the Aalborg University Model (AAU), has been particularly influential on the development of the methodology.

G. J. Costello, *The Teaching of Design and Innovation*, Contemporary Issues in Technology Education, https://doi.org/10.1007/978-3-030-41380-4_6

This chapter builds on the concept of design thinking (Curedale, 2013; Dorst, 2011; Dym et al., 2005) and enhances the approach by directly interfacing with industry/services professionals and entrepreneurs, while simulating a real-life interaction for the students. A further objective of the chapter is to contribute to the debate on the best pedagogical approach to developing undergraduate skills to meet the requirements of contemporary complex working environments. Furthermore, the cases will demonstrate how the concepts of phronesis, presented in Chap. 2; the concept of responsibility, presented in Chap. 3; and the concept of reflection, presented in Chap. 5, have been applied in a real-life educational context. The learning from the cases will be used to develop a generic process which will be presented in Chap. 7. An objective of this book is to contribute to practitioner education in the area of design and innovation. Consequently, a third industry-based case study is presented that provides an example of practitioner-researcher learning. This research project had significant influence on the teaching approach to design and innovation in the educational sphere and is also presented as an example of research-informed teaching.

6.1.1 The Value of Case Studies

Yin's (1994, p. 9) describes a case study as an "empirical enquiry that investigates a contemporary phenomenon within its real-life context" and where a "*how* or *why* question is being asked about a contemporary set of events over which the investigator has little or no control". Furthermore Leavy (1994, p. 105) concludes that case-based qualitative research now has an established pedigree in the field of business. Similarly Mårtensson and Lee (2004, p. 533) point out there is a growing acceptance of "intensive research methods" and that a single-site case study can be deemed scientific. Lee (1989) seminal paper also lays the foundation for single case studies which are also defended resolutely in the writings of Mintzberg (1979).

6.2 Case 1: Students Working with a Multi-national Company

The case described here was an academic class project carried out with Thermo King (2019), a multi-national subsidiary which was established in Galway, Ireland, in 1976, which presently employs over 500 people. The Thermo King Galway subsidiary is now the largest Ingersoll Rand plant in Europe and serves markets primarily in the Europe, Middle East, India and Africa (EMEIA) region in addition to supporting Latin America and Asia Pacific markets. Originally set up as a manufacturing location, it has recently added research and development (R&D) of the next generation of truck and trailer refrigeration units to its Galway capabilities. The opportunity to complete a project in a real-life product development environment, albeit in a controlled manner, had the potential to be a very beneficial learning experience for the final year students and help prepare them for their future careers. The results and analysis of the project are outlined in Table 6.1 using the following four constructs: positive aspects, recommendations for improvement, self-rating and needs assessment.

Now I will present comments by some of the students gathered during this review process to illustrate the results outlined in Table 6.1. Furthermore this approach allows the voice of the student to be adequately heard as is recommended in the AAU model (Jørgensen, 2004).

The first comment acknowledges the amount of effort that is required at the outset to develop the project and the commitment required by the students. However, the positive feedback from an external industry-based practitioner was very motivating for students.

> *A great deal of effort went into this project and to hear that they were impressed with our presentation gave me a great deal of satisfaction and made the project a worthwhile experience. Overall, I found that this project helped me gain a great deal of experience in product development and enabled me to improve my skills as an engineer.*
> *It is really helpful and motivating to talk with experienced professionals.*

The real-world scenario was appreciated by the students and working with the practitioners challenged them to "up their game". The realisation that their work

Table 6.1 Recurrent themes from case study 1

Construct	Results	Analysis
Positive aspects	Nearly all the students rated working on a real-life project (design brief) with professional engineers as the most positive aspect of the project. Also, the experience of working in a team was rated very highly	This result confirmed to the lecturer that the logistics of engaging with a company and adopting a flexible module road map was effective. It also was a motivation for the lecturer. Typically, such a project involves a lot of up-front work before module commencement
Recommendations for improvement	The most frequent recommendation was that the project should include the building of a prototype	Building of prototype has become easier and more cost-effective with the development of 3D printing. Motivation for lecturer to obtain funding for such a facility
Self-rating	The students were asked to rate their contribution to the project using the following taxonomy: *exceeded*, *strongly achieved*, *achieved*, *satisfactory* and *needs improvement*	Results would indicate that in the future the lecturer should correlate the self-rating of the student with actual module grades and provide feedback
Needs assessment	This has the widest categories of responses with suggestions for training on presentation skills, time management, market research, project management and research skills	This variety of the responses to this construct was not unexpected as self-development is very much based on the competence and skills of each student. However, this suggests that the lecturer should source some online individualised training for students

could have an impact was a source of satisfaction and a realisation that they were now "becoming engineers".

> *This project strongly simulated a real-world environment which is exactly what is wanted as a student of engineering.*
> *It was interesting to be part of the process in developing a product that has the potential of being introduced into the real world.*

The process addresses one area that is being increasingly identified as a priority for the education of engineers. Coming up with innovative ideas and technology is only one step. An engineer needs to be able to present and convince others that their engineering solution is worthwhile. A product is a lot more complex than just the technology it employs.

> *The presentation is another big positive as we had to present to an actual practitioner who was really interested in what we have to present. It is very satisfying as a designer to have people of their ability come in and be so enthusiastic about your project. [A] great sense of satisfaction.*

Each team project is assessed, and the same mark given to all students in a project team with 10% of the module marks for the presentation and 20% for the project report. Project assessment criteria involved inter alia the clarity of the presentation, the uniqueness of the solution and the feasibility of the solution, and, as it is an engineering module, the level of technical acumen was taken into consideration.

Other factors which are considered and may affect individual student's grade are attendance at weekly lecture/lab/team meetings, teamwork and contribution and construction of an artefact (alpha model) (Fig. 6.1).

6.3 Work Carried Out by the Students

Now I will briefly describe the process engaged in by the students with their real-life client to emphasise that the methodology simulates an industry-based project.

- This design brief was made available to the students via the draft design problem, and they prepared for a meeting with the client scheduled for the following week. The lecturer met with the class and presented an overview of the module learning outcomes and the structure of the project as well as assessment criteria and

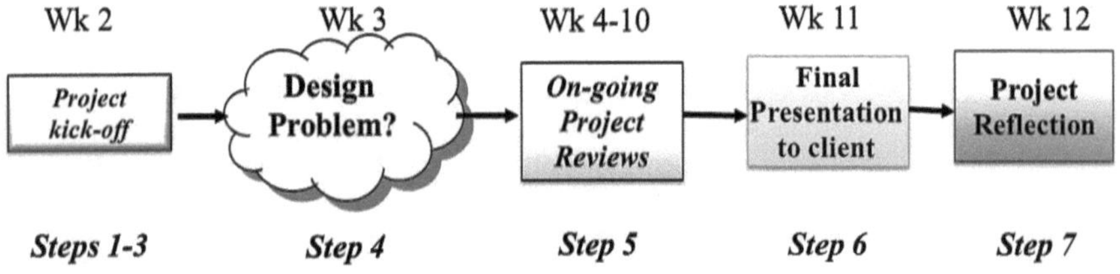

Fig. 6.1 High-level module road map. (Costello, 2014)

expected project logistics. Then the class was divided into project teams (normally three students per team).
- Together, the class project teams met the client face to face. The client presented the design problem to the class verbally with a more detailed description than in the design brief. This provided an opportunity for the class to get a more in-depth view of the client's thinking and to put themselves in the client's shoes (D. A. Leonard & Rayport, 1997). Also the project teams had time to question the client based on their initial week-long research into the problem domain.
- The project teams worked on the design problem during the semester using academic and industry standard product design methodologies (Cooper, 2001; Eppinger, 2001; Ulrich & Eppinger, 2004). The project teams completed a variety of tasks inter alia: a detailed project plan in the form of a Gantt chart; market research; customer needs analysis; developing and sketching design concepts; and ranking and choosing of the optimum solution. Each week the project teams presented a status of their work to the lecturer who in this type of pedagogy acts as a *coach* and *advisor* rather than the conventional lecturing mode. Agile development methodologies such as Scrum "sprints" were used to keep the project on track and to instil a discipline of reflection-in-action among the students.
- Towards the end of the semester, the class project teams presented their design solutions to the clients through oral presentation and a project report. These deliverables include a set of working drawings, computer-aided design (CAD) models, detailed target specifications, a human factors analysis of the proposed design solution, computer rendering of the proposed design and an artefact such as a mock-up of the design in cardboard or other materials.
- The feedback from the company was very positive, and they considered that the best design solution was good enough to introduce into their manufacturing process. This recognition was very motivating for the students.
- Each student had to complete a detailed project reflection that involved a report on their individual contribution to the project team. This was completed as a project in-class assessment worth 10% of the module marks. The focus on this end-of-project individual report was on how the student contributed to the team project. The idea for this assessment came from the lecturer's experience of having to complete end-of-year reviews while working in industry. Consequently, topics that the students were required to report on included the following:

  - Provide an overview of research **<u>you</u>** undertook in this project.
  - State the specific contribution that **<u>you</u>** made to the implementation of this project.
  - Describe what was innovative about your work as a member of the project team.
  - Identify your strengths and areas requiring development based on the responsibilities that you undertook.
  - Rate your performance:

Exceeded, strongly achieved, achieved, satisfactory, needs improvement and unsatisfactory.

If you choose exceeded, provide detailed justification.

- Briefly indicate what you consider were the positive aspects of the design and innovation project and any recommendations for improvement.
- Consent to use the information gathered for (confidential) research purposes (Yes or No).

However, students should be informed that there is some debate on the effectiveness of a single end-of-year performance appraisal. Critics argue that it is not agile enough for dynamic modern work environment and propose regular less formal reviews.

- The projects undertaken over a number of years have included the following examples: the design of a mechanical fixture to transport large components across Europe; a design brief for a media player case; a wetsuit drying apparatus; proposing products to utilise the extensive waste energy and materials from a craft brewery; developing a bicycle security lock utilising Internet of things (IoT) technology; and the design of a work station for a complex sub-assembly manufacturing process.

Now I will present a case where the students were involved in multidisciplinary projects.

6.4 Case 2: Students Working with Incubation Centre Start-Ups

In this case I will describe how the principles developed in Chaps. 2, 3, and 5 have been embedded in an educational module where cross-functional teams from engineering and business programmes in the Galway-Mayo Institute of Technology (GMIT) worked together on an entrepreneurial project. The project involved submission of a business idea to the Enterprise Ireland Student Awards competition (Student Awards, 2018). In the past 20 years, Enterprise Ireland has invested approximately €50 million in providing incubation centres to the third-level sector, located in both Universities and Institutes of Technology (Enterprise Ireland, 2018). The aim is to encourage the set-up of high-tech, knowledge-intensive enterprises. Enterprise Ireland funded two incubation centres in the Galway-Mayo Institute of Technology (GMIT) in late 2005 and mid-2006: one on the Galway campus and the other on the Mayo campus. The establishment of the Galway and Mayo incubation centres was part of an overall strategy of building regional innovation capability through Institutes of Technology. The Hubs provide start-up services and business development supports in a number of areas: financial, legal, sales and marketing, strategic planning, mentoring and networking (iHub, 2019). Enterprise Ireland organises the annual Student Entrepreneur Awards, which is open to all fulltime third-level students on the island of Ireland. The Student Entrepreneur Awards is part of an ongoing drive aimed at tapping into the potential for students to turn entrepreneurial ideas into commercial businesses. The objective is to encourage third-level students to start their own business as a career option. In 2018, the Mechanical and Industrial Engineering department and the Business School in GMIT collaborated with 11 cross-functional teams who submitted their ideas to the competition. The purpose of this case study is to demonstrate the process in which the teams operated and how it was designed to facilitate the students developing phronesis, responsibility and reflection – principles discussed in Chaps. 2, 3, 4, and 5 of this book. The five high-level steps in the process are outlined in Fig. 6.2. However, it must be remembered that the steps are for guidance rather than being prescriptive. The actual projects are iterative with multifaceted feedback loops and include many practical challenges that the team members must deal with, which require phronetic competences rather than being able to follow a rigid formula.

These practical challenges are outlined in Table 6.2, which I will refer to as the *Innovation Project Process*. The processes outlined in this case and the previous one were reflected on and developed further. This resulted in the SAL framework which will be described in Chap. 7.

Each step includes a commentary that explains how it contributed to development of learning in terms of phronesis, responsibility and reflection as discussed in earlier chapters.

A team has made it to the finals of the Enterprise Ireland all-Ireland competition (top 10) on two occasions over the last number of years. One proposed a product for use in the medical device industry and the other for use in the agricultural sector.

Despite calls to leverage the competencies of campus innovation centres (MacMahon et al., 2010), there is little evidence of actual collaboration in the literature. Furthermore, Mason and Arshed (2013) contend that "there is both little discussion in the literature on what experiential learning should take and a paucity of examples of experiential approaches to learning" (p. 449). This chapter proposes to address this deficiency in the body of knowledge of experiential learning. For example, reflection and feedback from the students is built into the module review process. As in case study 1 above, each student is required to do an assessment at the end of

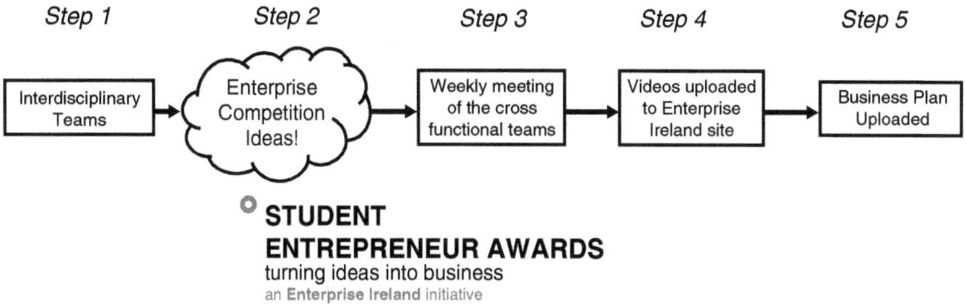

Fig. 6.2 Innovation team process

Table 6.2 Innovation project process

Step 1	A lecturer from both the Engineering and Business Schools met to organise student team members, agree relevant protocols, publish project strategies and examine the guidelines published by Enterprise Ireland that governed the 2018 competition
	Learning: *This meeting set the groundwork and deliverables that would facilitate the teams working and learning together. It also involved the lecturers working together within an ill-defined context. The ethical framework and reflective strategies were also discussed here*
Step 2	Project teams met, got to know each other and visited the student awards website in order to agree a shared meaning of the project requirements The teams then proceeded to the "ideation" stage in order to identify a business idea to meet the Enterprise Ireland criteria, i.e. technically advanced and with the potential to create jobs They also developed a detailed project plan in the form of a Gantt chart with deliverables and responsibilities
	Learning: *The ideation stage involved techniques such as brainstorming and mind-mapping as well as detailed research. Much of this work involved interactions and decisions that cannot easily be described and outcomes that cannot be predicted. Thrown together in a cross-functional team was a new and challenging experience*
Step 3	The teams presented their entrepreneurial ideas to the lecturers who were now acting as coaches and mentors rather than conventional lecturers Ethical aspects of their ideas had to be considered, and also the team had to consider any ethical issues associated with carrying out surveys or interviews. For example, one of the proposed projects involved improving access to trains for wheelchair users. The project team interviewed a mother of a wheelchair-bound child in order to empathise with the serious obstacles they have to overcome when undertaking a train journey Project teams developed a solution over the duration of the semester Project teams held weekly meetings during class where individual students had to take responsibility for chairing the meetings and taking/distributing minutes
	Learning: *Working together and communicating with colleagues from a different academic background involved making practical decisions within the context of new interpersonal relations. Lecturers no longer gave answers but advice and general direction. Decision-making was transferred to the students within the constraints of reconciling their disciplinary backgrounds and individual skills and competencies*
Step 4	The teams created a 3-minute video, specified by Enterprise Ireland, to demonstrate their idea focusing on the following criteria: (a) *What the product is* (b) *Who you think will buy it* (c) *How quickly you can commercialise it* (d) *Its export potential* (e) *Is there a competitor? If yes, why is your idea better?* The teams also presented their ideas to incubation hub managers who gave them salient feedback on the commercial potential of their ideas as well as their communication skills, e.g. how well they delivered an "elevator pitch"
	Learning: *The scripting and shooting of the video involved many decisions that the students would not have been previously been exposed to. Rules and regulations must be replaced by negotiation, motivation and persuasion. Something as imprecise as a video production involves practical decision-making requiring group consensus. Feedback from the Innovation Hub managers provided food for reflection*

(continued)

Table 6.2 (continued)

Step 5	The teams then proceeded to develop a business plan using the guidelines provided by the lecturers and a template available on the Enterprise Ireland student awards website As part of the review process, Enterprise Ireland competition judges choose 50 proposals from all of the videos submitted to the Enterprise Ireland website (in the order of 1000 submissions) Six of the GMIT cross-functional teams were successful in making it into the top 50. Their next step was to submit their business plan to the Enterprise Ireland site All the teams had to upload their completed business plan to the lecturers for grading via Moodle, an open-source learning management system supported by GMIT computer services
	Learning: *The development of the business plan required many prudent and practical decisions by the cross-functional teams. Many of the students described how they had learned from student team members from the other discipline. For example, the engineering students had to communicate technical aspects of their ideas so that they could be understood by the business students. The teams also had to navigate what for them is uncharted waters. For example, dealing with not being successful with Enterprise Ireland while keeping the team motivated to focus on getting the best grade possible*

the module. In this case the students had to present their contribution and learning verbally in a one-to-one interview with the lecture. The rationale used for this is based on the author's experience (20 years as an engineering and product management practitioner) of having to complete end-of-year face-to-face reviews. Furthermore, it challenged the students to develop verbal skills in this simulated interview situation. I remind the students that the next time they will have to verbally describe the project could be at a job interview, and this exercise was designed to prepare them for this.

The feedback from this process is important for the lecturer who is continually endeavouring to reflect on and improve the module content and process year on year.

6.5 Ethical Clearance

Data gathering for the research was completed by means of the students providing feedback on the module using a structured template. Students were asked to give or decline their assent on using the data for research purposes. Furthermore the students were given assurance that any data would be anonymous. Permission from the practitioners to use their interaction with the project for research purposes was obtained by email.

This case is an attempt to address the dearth of "practical" empirical studies on the teaching of phronesis. The objective of the cases was to facilitate the learning of phronesis, responsibility and reflection in educational practice. I would also suggest that much of the conclusions of this study is applicable to professional managerial education and practice. Consequently, I will present a case study of educating a practitioner to design and develop an innovative organisation.

6.6 Case 3: Industry-Based Project as Research-Informed Teaching

Now I will proceed to outline a case study of action learning in the context of a practitioner working on an innovation management project in his own organisation (Costello, Donnellan, & Conboy, 2013). Furthermore, the case is presented as an example of research-informed teaching where the researcher took learning from the project "into the classroom". The study began in APC Ireland, a subsidiary of the American Power Conversion (APC) Corporation. The corporation entered a major period of transition in the first quarter of 2007 with completion of its acquisition by Schneider Electric and the formation of a new subsidiary called APC (by Schneider Electric). The explosive growth of the internet has resulted in the company broadening its product offerings from uninterruptible power supplies (UPS) to the high-end InfraStruXure™ architecture in order to meet the critical availability requirements of internet service providers (ISP) and data centres. APC had two locations in the west of Ireland that served Europe, the Middle East and Africa (EMEA). The manufacturing operations site employed approximately 100 people, while a number of functions on the second site, including sales, information technology, business support and software research and development, had a workforce of approximately 300. Responding to the supply chain challenge, a Lean Transformation project was set up in the manufacturing location with a cross-functional team of 12 members drawn from management, engineering, manufacturing, materials planning, quality and logistics functions (Womack & Jones, 2003, 2005). The Lean Transformation Team set an objective to quickly deliver the message that APC Ireland was responding to, and leading, the corporate initiative while also providing a platform for the Irish subsidiary to obtain a reputation as an innovative location. The manager of the manufacturing site began collaborating with the researcher (the author of this book).

6.6.1 Phase One of the Project

This first phase of the project lasted 1 year. Data collection methods during this phase involved maintaining a log book, reviewing documents and information systems, compiling records, conducting interviews, recording observations (direct and participant), studying artefacts and carrying out surveys in order to develop a database and body of evidence (Gillham, 2000; Yin, 1994). A total of 29 unstructured or "open" interviews were undertaken that involved approximately 60 hours of interview time and 24 days spent observing the company sites. The interviews were conducted across a wide area of the organisation, including with senior managers with global, EMEA (Europe, Middle East and Africa) and Irish site responsibilities, middle managers, team leaders, engineers and a number of people in general planning roles. A main focus of this phase was the "Lean" project described above. Furthermore, the researcher had the status of a temporary employee with his own email address and intranet access.

6.6.2 Phase Two of the Project

The decision to move to an action research cycle emerged from the phase one interaction and was based on the ambition of the manufacturing subsidiary to become an innovative location. It was realised that such a project would involve significant change within the organisation and its processes. Resulting from discussion with the plant manager, there was an agreement to move forward using a dialogical AR approach with researcher-plant manager meetings every 2 weeks. As outlined in Chap. 5, Mårtensson and Lee (2004, p. 531) propose that "reflective dialogues outside the organisation can help the manager to reflect on, learn from, and remedy managerial problems in the organisation". Data collection during this period involved recording of the meetings which were subsequently transcribed verbatim by the researcher. Given the rich nature of the data, this was considered the optimum way of capturing the reflective meaning and ensuring consistent interpretation. Analysis was done manually through the examination of each meeting transcript and through providing a summary of the topics discussed in the transcripts. This then was sent to the plant manager for his evaluation and confirmation that it was an accurate portrayal of the meeting, as advocated by Kelly and Murnane (2005). In total these transcripts ran to over 60,000 words.

6.6.3 Reflection on the Dialogical AR Project

This section will provide a reflection on the dialogical AR project guided by literature contributions in the areas of innovation (Tidd, Bessant, & Pavitt, 2005), organisational change (Tushman & O'Reilly, 2004) and leadership (Nadler & Tushman, 2004). On the first examination of the study, the evidence suggests that the organisation was undergoing a process of adaptation since the initial "Lean" project was undertaken reactively in response to APC corporate communications that there was a need for improvement in process innovation (delivery of products and services) and paradigm innovation (organisational models). Subsequent to the acquisition by Schneider Electric, the manufacturing operations subsidiary quickly embraced the principles of the Schneider Production System (SPS), which is closely related to the Lean approach of the Toyota Production System (TPS). This involved visiting a flagship plant in France that uses SPS and networking with some of the main corporate leaders and implementers of the program. An example of this was the running of a major Kaizen event guided by Schneider's experts in the area, which was a first for any of the APC subsidiaries. Kaizen, a Japanese word for improvement that has become associated with Lean practices, is a process improvement approach that is integral to Lean thinking. It is interesting for this study that Tidd et al. (2005) propose the practice of Kaizen as a method of continuous incremental innovation over a long period. A major process innovation was introduced to the plant, namely, short interval management (SIM). The method was implemented in the APC operations site both as a communications instrument and as a tool to help with the running of a production line. SIM is used to communicate issues from the line up through the organisational support structure so that they can be prioritised and addressed.

The plant manager had this reflection on the SIM implementation:

> The best way to get good ideas is to get lots of ideas. In terms of our organisational change, the SIM process has put a mechanism in place that allows people to get their ideas implemented. While the majority might be small and incremental – bigger ideas can emerge. For example, the SIM process threw up a potential problem with our health and safety process – it was too dependent on one person. The result was that we implemented an organisational change – and the external auditors were so impressed by the SIM process contribution to H&S [health and safety] that we won a national award. People are inherently intelligent, but you need a mechanism to allow people to use their intelligence. The SIM process now facilitates people using their natural creativity and make suggestions that will be implemented. We didn't have this before and, we are keeping a database of the suggestions.

While from a day-to-day perspective, the incremental and reactive nature of the process of organisational change is most obvious, the dialogical AR project consisted of both strategic and anticipatory aspects. The vision of the plant manager was that manufacturing operations be transformed from being "a manufacturing subsidiary" that produces a certain product line to an "innovative subsidiary" that can adapt to changing business process requirements and hence to be a sustainable location that can accommodate the exigencies of a continually evolving corporate product portfolio and environment. When asked what motivated him to undertake the organisational change process to implement an innovation culture, he had this reflection:

> I saw it as a means to engage people and get them excited about something; start using their minds at work and overall making the site a better place to work in.

This reorientation includes a long-term perspective that encompasses strategic change, the building of networks (both within the corporate organisation and externally with academic and other sectors) and, most importantly, the creation of an environment that allows and encourages every person to be innovative and effective:

> For me, if I reflect on the last twelve months, I would say that before we started this [project] an innovative culture was something that you could almost not define: it was just an airy-fairy type of concept. After having gone through the process over the last while, being able to define a structure that helps support an innovative culture has been a key point.

The incumbent leadership approach consisted of three dominant behaviours: structuring, controlling and rewarding. In the case of structuring, there has been a concerted effort to building and empowering teams as evidenced by the "Lean Transformation Team", which received corporate recognition. In relation to this, the concept of "empathy" emerged in the discussion in the following way:

> The reality of working in a fast-changing supply chain environment demands that people have the human quality and attitude of being able to cope with imperfect or messy processes and deliver for the team.

The manufacturing environment and the emphasis on measurement and key performance indicators (KPIs) in the SIM process naturally support the controlling attribute. Furthermore, within the limited budget constraints, there has been an attempt to provide some rewards to employees and teams with the SIM process providing a quick feedback loop to ensure corrective actions do happen and responsibilities and commitments are adhered to. That said, the changing environment and the concerns about the future are real issues. The plant manager commented:

> That creates a problem for what we are doing here because for people to be creative and innovative, your morale needs to be high. If the morale is going to be low, people are not going to come in and be creative. They will be more concerned about their jobs, so you need a secure environment [for innovation].

Having completed the reflection on the project, we will now discuss learning outcomes from the project for both the practitioner and researcher and how this became research-informed teaching of design and innovation.

6.6.4 Learning from the Dialogical AR Innovation Project

This section will outline learning from the industry-based case study that informed the teaching approach of the researcher in his capacity as a lecturer. This is presented as an example of "research-informed teaching", and after each learning example, I will indicate how it influenced the teaching practice by presenting "teaching points".

A significant and new process innovation, short interval management (SIM), was introduced during the AR study, and this became the focus of the practitioner and researcher attention during phase two of the project. Using the broad definition of an information system proposed by authors such as Verrijin-Stuart (1989), Whitten, Bentley, and Ho (1986) and Fitzgerald, Russo, and Stolterman (2002), SIM is a de facto information system which acts as the human interface to the enterprise resource planning (ERP) system. Furthermore, this transformation was likened to a digital-to-analogue conversion process by the plant manager. The human activity results in the ERP information which are then prioritised and acted on. In this conceptualisation the SIM becomes an interface and a transformation location, between the digital world where the information resides and an analogue world where the information is acted on and implemented.

Teaching Point: *This finding influenced my design of the "Practicum" in the next chapter and its conceptualization as a transformative location. Here the students would synthesise the digital world where information was located and developed (such as CAD software), and its implementation in a real world where people relate in an "analogue" way. An overt reliance and fixation with technology in the digital age needs to be constantly questioned and guarded against. This is particularly relevant in a design and innovation environment that aims to instil human qualities such as empathy.*

The chief communication vehicle of the process is the SIM "cell information board" where the health and safety, quality and production issues are captured and dealt with in that priority. To enforce the priority given to health and safety, the supervisor is responsible for posting a photograph of the problem on the board. The board information has a standard format with the health and safety issues on the extreme left, followed by quality issues and production issues with the priority from left to right. Any potential barriers to meeting the build plan are identified and if necessary escalated to the support staff. The cell support team holds a mandatory daily meeting which should be a maximum of 30 minutes which is run by the production manager. Membership of this team includes the manufacturing and quality engineers, the material specialist and the cell supervisor. These meetings are also open to any of the management team to attend. The agenda follows the priority list referred to earlier where the following issues are reviewed: health and safety, customer feedback, quality issues, issues escalated from the production SIM meeting, agreed action items from the last meeting, feedback from operations personnel on active issues and finally agreed issues for action prior to the next meeting. The last item requires actions to be both specific and deliverable with the focus on meeting the short interval rather than any long-term requirements. The layout of the SIM cell information board, shown in Fig. 6.3, reinforces the priorities outlined above.

Teaching Point: *There is a dearth of literature on how to develop reflection-in-action among students. This example suggested the need to provide regular stand-up meetings with the students focusing on the important issues in the project. To address this, I experimented with the use of "agile" methodologies involving stand-up "sprints" with individual teams at the beginning of class. My classroom whiteboard became my version of a CIB. I also specifically included "reflection" as a unique component of the "sprint" meeting.*

While there have been regular calls for more effective academic and practitioner collaboration, evidence suggests that this is still largely unfulfilled (Ågerfalk, 2010; I. Benbasat & R. W. Zmud, 1999; Hevner & Chatterjee, 2010; Schön, 1990). The plant manager commented at lunch that when this project had started, he thought of it as just a nice thing to do:

> However, after the visit of the Schneider VP, I saw the value of what we were doing. If we hadn't been able to talk about the research, we would have only been another manufacturing plant. I then realised the benefit of the research and of publications.

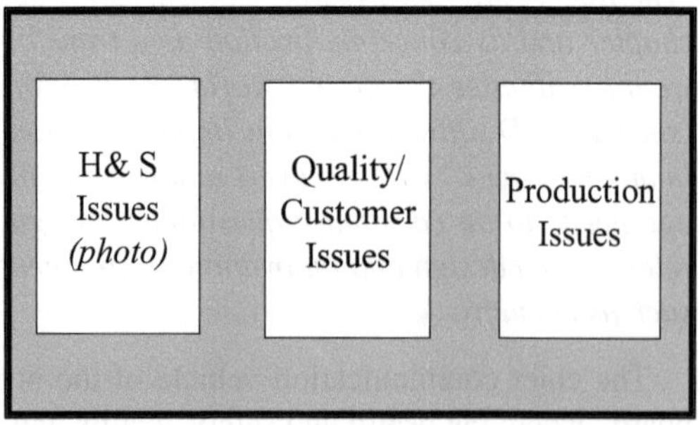

Fig. 6.3 Cell information board (CIB)

The structure of this research had a very significant practitioner commitment that even involved joint presentations of the work to academics. I believe that such a dual focus can assist other practitioners and academics, both from the context and "commonalities" emerging from the study and from the voice of a practitioner on how research can contribute to a dynamic real-world situation. The approach is also consistent with Susman and Evered's (1978) argument that action research – and in this case dialogical AR – assists members of the organisation develop their capability to resolve their own problems. Here is the plant manager on this topic of empowerment:

> [In] the present manufacturing environment, people come to the plant to work and then go home. I want people to come in everyday to work and think – and then to go home. I want to tap the resources of the people and create a habit of learning.

Teaching Point: *The theme I gleaned from these comments was the importance of meaningful interactions for the students with working professionals. This was done with a visit to the practitioners work environment where the students detailed overview of the context for the design and innovation challenge. Following this, students were tasked with providing an interim communication of their design concepts and a final presentation in a suitable location in the college. Furthermore, the comment on the importance of empowerment led me to coach the teams to tap into their own resources as this competence would be required when they began to work in a real-life environment where they would have significant responsibilities.*

The advantage of dialogical AR is that the reflective one-to-one dialogues inherent in the approach involve regular opportunities to engage with, and reflect on, the *process of reflection.* The plant manager states:

> There is real benefit by bringing the literature to me. … People from the academic world read lots – that's what they do. However, being able to filter it down to the likes of me who might want to read but does not have all that time is important.

An example of the evolution in the practitioner's reflective experience was the development of his understanding of the meaning of the term "innovation culture":

> A significant output for me was just the whole learning process of putting a structure around an innovative culture – it is not just a fuzzy thing – I have a clear image in my mind of what that means.

And another example from the plant manager of the maturing of his thinking about innovation is described as follows:

> If another manager came to me and said, 'I want to be more innovative – what does that mean?' I could advise them and share a lot of this [project] with them which would be significant – a framework to go forward that I would intend to apply to my day-to-day work.

A key result from the scheduled evaluations at the end of each stage of the AR cycles was that when the reflection was carried out in an ad hoc manner, it had little impact on the practitioner. However, when a structured questionnaire was used that was designed to stimulate the reflective process, the practitioner described it as being very beneficial to his process of learning:

> I see a great value in this research by forcing me to take time out for reflection.

Teaching Point: *The process of reflection needs to be developed and ingrained in the cognition of inexperienced students. This led me to "require" each team to take time for reflection during the class stand-up meetings with the object that it became deep-rooted in their thinking.*

The dialogical AR provided an interpretive space for the practitioner. The importance of this factor for innovation has been emphasised by Lester and Piore (2004) to compliment the analysis dimension of innovation.

Teaching Point: *The regular tiered college classroom is not conducive to operate as "interpretive space". The setting for this type of teaching requires facilities that allow and encourage interaction. However, this can be a challenge in schools with limited resources and may require innovative solutions,* e.g. *using available space in the college for teams to meet such as the canteen at nonpeak times.*

In this project the practitioner found it very beneficial that the researcher had summarised academic literature that he thought would be of value to the situation under review. This would be in line with a dyadic view, that is, when one member of a dyad develops, then the other develops as well.

One of the feedback items was that the project and interactions provided an intellectual stimulation for the practitioner that would not normally be associated with the world of praxis.

Teaching Point: *Many studies have shown that intrinsic motivation is the result of working on an interesting challenge and developing a solution that is recognised to be of a high standard. In case study 1, the practitioners gave nominal prizes to the teams with the best overall design solution, most innovative idea and best presentation. This was observed to be a stimulating impetus for the students to achieve excellence in their work. The teacher/coach must adopt the role of "walking with the students" to optimise the dyadic nature of this type of teaching.*

Van de Ven (2010) highlights the need for *humility* when carrying out Engaged Scholarship. The author would at this stage broaden this point by adding the concept of *humbling*. The enthusiasm, openness and accessibility of the interviewees were a genuine humbling experience for the researcher.

Teaching Point: *Working with student teams in this capacity requires a certain "humility" by the teacher/coach. The intensive meetings with the teams need to be approached using the insight from Erasmus quoted in Chap. 1: that education must "draw out" design and innovation from the students. The dissimilar role is a challenge for a lecturer who may need to develop new skills and strategies for the interaction with the students.*

Van de Ven also makes the point that engagement is a *relationship* where practitioners and scholars co-produce knowledge. The researcher built up a "friendship" with many of the interviewees which has remained long after the research was completed.

Several academic papers were published during the ongoing research. The practitioners were included as co-authors in order to do justice to their input. This was also a source of motivation and satisfaction for the practitioners who saw their worked recognised and included in the academic discourse.

Teaching Point: *Writing up the case by the teacher/coach and presenting academic papers is a potential source of motivation and opportunity for reflective learning. The literature suggests that teaching design and innovation is nascent area that requires more research and debate. This case study has demonstrated that research-informed teaching has the potential to make significant academic contributions.*

This case study provides an example of how practitioner-academic reflective dialogues can provide positive business and learning results. Furthermore, the engagement of the author in the research provided a learning experience that he was able to transfer to the classroom as research-informed teaching. Such interactions involve practical judgement that defies deontological series of rules. It requires phronetic decisions made in the context of the reflective dialogues.

6.7 Conclusions

This chapter provides an example of using student-practitioner collaboration in the teaching of design and innovation modules to mechanical engineering students, which also involved collaborating with business students. There were several learning experiences in this study: by the students, by the lecturer and by the industry partner. Additionally, the act of writing up this study provided a reflective learning experience for the author. The module structure, described here, has embedded entrepreneurial learning in the GMIT department of Mechanical/Industrial Engineering. Working directly with industry practitioners and entrepreneurs is a novel pedagogical approach that fosters entrepreneurial thinking and behaviour among the students. Furthermore, key stakeholders (in this case the multi-national project managers and staff of the Innovation Hubs) have been persuaded to engage in the learning process. These managers have been very supportive of the process as it meets their remit to involve both industry and incubation centres with academia. The lecturer sought feedback from the practitioners on the learning process and how future projects could be improved. Reaction to the projects was positive as the students appreciated the opportunity to work in a *simulated* environment analogous to industry. Students were particularly pleased that their work might be implemented in a real-world product and not just be archived as another class project. The current literature largely associates simulation-based learning to computer applications such as games. However, this chapter disagrees with technology-focused opinions such as by Chang, Chen, Lin, and Sung (2008), who identify SBL as solely computer-based. In contrast, it argues that role-playing by students interfacing with companies and other organisations such as technology start-ups can also be regarded as "simulation" in a wider sense.

This study is set against the background painted by Gavin (2011).

> Engineering education is in a state of flux, with universities facing requirements from industry to develop graduates with a wider skills base, while at the same time a revolution in the availability of information is changing the way that students learn (p. 547).

To address this, I propose the concept of *simulation-action learning* (SAL) for students as an enhancement of problem-based learning based on the empirical work presented in this chapter that is informed by the "principles" presented in Chaps. 2, 3, 4, and 5. The next chapter will outline SAL in detail. It also contributes to the what Gattieab et al. (2011, p. 521) call the cultivation of education "as a complex system that will prepare students to think critically and make decisions with regard to poorly understood, ill-structured issues".

Chapter Summary

- Three case studies were presented as examples of using the principles outlined in the previous chapters in simulated real-world problems.
- Case 1 described a design and innovation project with a multi-national company where the students were immersed in a real-life design brief simulating the work of design engineers.
- Case 2 described a design and innovation project that involved cross-functional teams of engineering and business students within the structure of a national enterprise competition.
- Case 3 described an innovation research project based in industry. The learning for the practitioner involved frequent one-to-one reflective dialogues and is offered as an exemplar for continuous professional development (CPD) courses in design and innovation.
- The engagement of the author in the research project resulted in the transfer of this learning to educational practice where the researcher was able to bring the learning into the classroom as research-informed teaching. In addition, it enabled implementation of the principles outline in the earlier chapters.
- The overall strategy in all cases was immersion and simulation of real-life scenarios.
- These themes will be developed into a theoretical framework in Chap. 7.

References

Ågerfalk, P. J. (2010). Editorial: Getting pragmatic. *European Journal of Information Systems, 19*, 251–256. https://doi.org/10.1057/ejis.2010.1022

Benbasat, I., & Zmud, R. W. (1999). Empirical research in information systems: The practice of relevance. *MIS Quarterly, 23*(1), 3–16.

Boud, D., & Feletti, G. (1998). Changing problem-based learning. Introduction to the second edition. In D. Boud & G. I. Feletti (Eds.), *The challenge of problem-based learning* (pp. 1–14). London: Kogan.

Chang, K.-E., Chen, Y.-L., Lin, H.-Y., & Sung, Y.-T. (2008). Effects of learning support in simulation-based physics learning. *Computers & Education, 51*, 1486–1498.
Cooper, R. G. (2001). *Winning at new products: Accelerating the process from idea to launch*. New York: Perseus Publishing.
Costello, G. J. (2014). *Teaching product design through industry collaboration*. Paper presented at the Proceedings of the ASME 2014 12th Biennial Conference on Engineering Systems Design and Analysis ESDA2014 Volume 1, June 25–27, 2014, Copenhagen, Denmark.
Costello, G. J. (2016). Incubation centres and the teaching of innovation: Bridging theory and practice. *Journal of the All-Ireland Society for Higher Education AISHE-J, 8*(1). Available online through http://ojs.aishe.org/index.php/aishe-j/article/view/242
Costello, G. J., Donnellan, B., & Conboy, K. (2013). Implementing an innovation project in an Irish multi-national subsidiary: A dialogical action research study. *Irish Journal of Management, 33*(1), 73–89.
Curedale, R. (2013). *Design thinking: Process and methods manual*. Design Community College. Topanga, California.
Dorst, K. (2011). The core of 'design thinking' and its application. *Design Studies, 32*(6), 521–532.
Dym, C. L., Agogino, A. M., Eris, O., Frey, D. D., & Leifer, L. J. (2005). Engineering design thinking, teaching, and learning. *Journal of Engineering Education, 94*(1), 103–120.
Engineers Ireland. (2019). *Engineers Ireland website*. http://www.engineersireland.ie. Accessed July 2019.
Enterprise Ireland. (2018). *Enterprise Ireland website*. http://www.enterprise-ireland.com/en/. Accessed April 2018.
Eppinger, S. D. (2001). Innovation at the speed of information. *Harvard Business Review, 79*(1), 149–158.
Fitzgerald, B., Russo, N., & Stolterman, E. (2002). *Information systems development: Methods in action*. Berkshire, UK: McGraw-Hill.
Gattie, D. K., Kellam, N. N., Schramski, J. R., & Walther, J. (2011). Engineering education as a complex system. *European Journal of Engineering Education, 36*(6), 521–535.
Gavin, K. (2011). Case study of a project-based learning course in civil engineering design. *European Journal of Engineering Education, 36*(6), 547–558.
Gillham, B. (2000). *Case study research methods*. London: Continuum.
Hevner, A., & Chatterjee, S. (2010). *Design research in information systems: Theory and practice (integrated series in information systems)*. New York/London: Springer.
iHub. (2019). *GMIT innovation hubs*. https://www.gmit.ie/gmit-innovation-hubs. Accessed July 2019.
Jørgensen, F. (2004). The student voice. In A. Kolmos, F. K. Fink, & L. Krogh (Eds.), *The Aalborg PBL model: Progress, diversity and challenges* (pp. 21–35). Aalborg, Denmark: Aalborg University Press.
Kelly, S., & Murnane, S. (2005). Academic performance evaluation and the organisation of knowledge in the research-intensive university. In N. Harvey (Ed.), *The Irish Journal of Management incorporating IBAR: Selected papers from the 2005 Irish Academy of Management Conference* (Vol. 27, pp. 95–109). Dublin, Ireland: Blackhall Publishing.
Lee, A. (1989). A Scientific Methodology for MIS Case Studies. MIS Quarterly March 1989, 33–50.
Leavy, B. (1994). The craft of case-based qualitative research. *IBAR - Irish Business and Administrative Research, 15*, 105–118.
Leonard, D. A., & Rayport, J. (1997). Spark innovation through empathic design. *Harvard Business Review, 75*(6 November–December), 102–113.
Lester, R. K., & Piore, M. J. (2004). *Innovation-the missing dimension*. Massachusetts: Harvard University Press.
MacMahon, C., Coleman, M., Ledwith, C., Cliffe, B., & McGlone, R. (2010). *Accelerating Campus Entrepreneurship (ACE): A Sectional Analysis of Practices to Embed Entrepreneurship Education into Engineering at Irish Higher Education Institutions*. Paper presented at the ISEE 3rd International Symposium for Engineering Education 1-2 July 2010.

Mårtensson, P., & Lee, A. S. (2004). Dialogical action research at omega corporation. *MIS Quarterly, 28*(3), 507–536.

Mason, C., & Arshed, N. (2013). Teaching entrepreneurship to university students through experiential learning. *Industry and Higher Education, 27*(6), 449–463.

Mintzberg, H. (1979). An emerging strategy of "direct" research. *Administrative Science Quarterly, 24*(4), 582–589.

Nadler, D. N., & Tushman, M. L. (2004). Beyond the charismatic leader: Leadership and organizational change. In R. Katz (Ed.), *The human side of managing technological innovation: A collection of readings* (2nd ed.). Oxford, UK: Oxford University Press.

Savin-Baden, M. (2003). *Facilitating problem-based learning: Illuminating perspectives*. Maidenhead, UK: Society for Research into Higher Education & Open University Press.

Schön, D. A. (1990). *Educating the reflective practitioner: Toward a new design for teaching and learning in the professions*. San Francisco: Wiley.

Student Awards. (2018). *Student entrepreneur awards*. Available on-line at http://studententrepreneurawards.com/. Accessed April 2018.

Susman, G. I., & Evered, R. D. (1978). An assessment of the scientific merits of action research. *Administrative Science Quarterly, 23*(4), 582–603.

Thermo King. (2019). *Thermo King website*. https://www.thermoking.com/na/en.html. Accessed July 2019.

Tidd, J., Bessant, J., & Pavitt, K. (2005). *Managing innovation: Integrating technological, market and organizational change*. Chichester, UK: Wiley.

Tushman, M. L., & O'Reilly, C. (2004). The ambidextrous organization: Managing evolutionary and revolutionary change. In M. L. Tushman & P. Anderson (Eds.), *Managing strategic innovation and change: A collection of readings* (2nd ed.). Oxford, UK: Oxford University Press.

Ulrich, K. T., & Eppinger, S. D. (2004). *Product design and development* (3rd ed.). Boston/London: Irwin McGraw-Hill.

Van de Ven A. H. (2010). *Reflections on engaged scholarship (paper delivered to Carlson School of Management, University of Minnesota)*.

Verrijin-Stuart, A. (1989). Some reflections on the Namur conference on information system concepts. In E. Falkenberg & P. Lindgreen (Eds.), *Information systems concepts: An in-depth analysis*. North Holland: IFIP.

Whitten, J. L., Bentley, L. D., & Ho, T. I. M. (1986). *Systems analysis & design methods* (2nd ed.). Santa Clara, CA: Times Mirror/Mosby College.

Womack, J. P., & Jones, D. T. (2003). *Lean thinking: Banish waste and create wealth in your corporation*. London: Free Press.

Womack, J. P., & Jones, D. T. (2005). *Lean solutions*. New York: Simon & Schuster.

Yin, R. K. (1994). *Case study research: Design and methods*. London: Sage Publications.

Chapter 7
Simulation-Action Learning (SAL)

7.1 Introduction: Rationale for the Proposed Framework

Ruse (2005) defines theory as an "attempt to bind together in a systematic fashion the knowledge that one has of some particular aspect of the world experience". This chapter binds together the world experiences, described in Chap. 6, into a theoretical framework called simulation-action learning (SAL). The SAL phased process is set out in detail to assist teachers in building modules for design and innovation that simulate real-work conditions. The process can be applied to a wide variety of student cohorts, and, in addition, SAL can be adapted to provide short continuing professional development (CPD) courses. To provide a background, the chapter will first trace the development of problem-based (or project-based) learning and its progeny simulation-based learning, as pedagogical approaches. There can be a certain amount of confusion with the terms *problem-based* and *project-based* learning. For the purpose of this book, I will take it to be a matter of duration: problem-based tasks are of short duration (a single class or over a few weeks), while project-based assignments extend over a semester or year. The chapter will utilise Donald Schön's concept of a *practicum*, a purposefully designed setting in which practice can be simulated to approximate a real-world scenario. This concept will be a critical phase in the SAL process. The initial review of PBL is focused on the engineering discipline, but its application is broadening, as shown in case study two in Chap. 6, which involved engineering and business students collaborating on projects.

The application of problem (or project)-based learning (PBL) in an educational context is frequently traced back to the Danish tradition established in two universities in the 1970s. One of these, the Aalborg University (AAU) model, has been particularly influential on the development of the methodology. The theoretical basis is founded on the works of Piaget, Dewey, Lewin and more recently Kolb (Kolmos, Fink, & Krogh, 2004). The approach is described as concerning "three dimensions: the problem, the content and the team" (ibid. p. 12). An important part of the development of the model has been the inclu-

G. J. Costello, *The Teaching of Design and Innovation*, Contemporary Issues in Technology Education, https://doi.org/10.1007/978-3-030-41380-4_7

sion of "the student voice" (Jørgensen, 2004). Furthermore the progress of the AAU is seen as a response to societal change and its implications for education (Krogh & Rasmussen, 2004) as well as the "new global realities" which are "at best described as 'complex' and 'uncertain'" by Sørensen (2004 p. 110). According to Luryi et al. (2007), engineering programmes increasingly endeavour to include entrepreneurship and innovation in the curriculum. There is a significant literature on problem-based learning in engineering education (Perrenet, Bouhuijs, & Smits, 2000). The focus in this type of learning is to provide the students with problem scenarios so that they can learn through a process of action and reflection. The environment of engineering has radically changed in the last decade driven by advances in information and communications technology. Furthermore, globalisation of manufacturing and R&D (research and development) has had a significant impact on how engineers work. Among their recommendations is that engineering programmes "should involve hand-on business experience based on innovating engineering projects" (Luryi et al. p. T2E-15). Their study included a review of two major journals in the area of engineering education using the search word *entrepreneurship* that yielded the following results. The *Journal of Engineering Education* (JEE) had 14 publications about entrepreneurship from 2001 to 2009, while the *European Journal of Engineering Education* (EJEE) had 10 publications about entrepreneurship from 2000 to 2012. Examples from the former journal include:

- Ohland, Frillman, Zhang, Brawner, and Miller III (2004) who conclude that entrepreneurship programmes are beneficial to engineering students
- Creed, Suuberg, and Crawford (2002) who argue for a paradigm shift that requires the merger of classroom learning and industry participation
- Mendelson (2001) who proposes joint projects between engineering and business students

EJEE publications include studies by:

- Silva, Henriques, and Carvalho (2009) who argue that teaching product development in an entrepreneurship framework promotes students' skills
- Papayannakis, Kastelli, Damigos, and Mavrotas (2008) who contend that entrepreneurship teaching should be part of a more general discussion related to educational priorities
- Casar (2000) who proposes a synergy between research and education

These publications support the argument of this chapter that direct collaboration between a practitioner (such as an entrepreneur) and students has a strong pedagogical basis that goes beyond the traditional PBL model (Tan & Ng, 2006). Furthermore it contributes to what Lappalainen (2011) terms the "ability for critical engagement and thought, interdisciplinary and original thinking, collaborative teamwork, and socialisation into the engineering community" (p. 513). Now, I will review the concept of simulation as part of the learning process and argue that it is relevant to design and innovation pedagogy.

7.2 Simulation and Learning

The Oxford Dictionary defines simulation as follows: to "imitate the appearance or character of" (ODE, 2006). While the entry does refer to computer modelling as an application of simulation, I will use the broader definition to argue that the etymology of the word implies a much wider concept than that of computer modelling and the digital learning debates (Eck, 2006; Prensky, 2001).

Recently there has been a body of literature in the field of medicine on the subject of using simulation-based learning (SBL) as an enhancement of PBL (Cant & Cooper, 2010; Lateef, 2010; Steadman et al., 2006). PBL had its origins in 1968 in a medical programme at McMaster University in Canada and subsequently was adopted in other disciplines such as engineering (K. A. Smith, Sheppard, Johnson, & Johnson, 2005). Its influence on medical education and training is supported by the fact that Stanford has a centre dedicated to its study (CISL, 2019). Simulation is being used to increase nurse's self-efficacy and skills (Fadale, Tucker, Dungan, & Sabol, 2014), while a review of simulation-based learning by Cant and Cooper (2010 p. 3) concludes that simulation "using manikins is an effective teaching and learning method when best practice guidelines are adhered to". In the area of surgery, simulation-based learning models attempt to replicate an environment similar to real-life surgical situation (Khunger & Kathuria, 2016). Importantly for my argument, Lateef (2010) emphasises that simulation-based learning is a *technique* not a *technology* and is designed to "replace and amplify real experiences with guided ones, often 'immersive' in nature, that evoke or replicate substantial aspects of the real world in a fully interactive fashion". Consequently, this chapter proposes to contribute by transferring the simulation concept from medicine to engineering education in a similar way that PBL migrated from medicine to engineering.

The topic of simulation as a learning experience has emerged in the management literature, where, for example, Lu, Hallinger, and Showanasai (2014) found that students considered SBL to provide a richer learning experience than conventional methods such as lectures. They also proposed that SBL addressed the increasing criticism with the management literature on the relevance of much educational pedagogy in the field. In the area of teaching entrepreneurship to management students, the role of the teacher changes from that of a presenter to that of a "coach" (Cadotte, 2014). Deegan et al. (2014) in their paper on the role of simulation-based learning environments (SBLE) in public management curricula proposed out that it addresses the genre of "wicked problems" and complexity which Lindblom (1959) had examined a few decades ago.

In the teaching of design, Davis (2017 p. 191) describes simulation as "an imitation of a real-world process or system behaviour over time". In this regard, the SAL process, presented later in this chapter, is completed over an extended period. Furthermore, she contends that the critical and creative use of simulation helps "refine ideas for, and provide specifications for, a real-world working solution, rather than a theoretical one" (ibid.). In the area of architectural design, Sara (2011 p. 9) defines a live project as a "type of learning which is distinct in its engagement of real 'external collaborators' such as clients or users for a particular piece of student work". These types of projects sit between "the binaries of theory and practice, university and community, designing and making, the head and the hand, and ideas about what it to be a student, what it is to be a professional" (ibid.). She contends that the skills and learning involved in live projects are not discipline specific and are transferable to other fields. Her suggestion that live projects can create new knowledge and approaches to professional practice resonates with the aim of this chapter to develop a conceptual framework for the teaching of design ad innovation.

However any review of the literature must be cognisant of the assertion by Cooney and Murray (2008 p. 19) that the debate continues on whether or not design and innovation, together with associated competences such as entrepreneurship, can even be taught.

In Chap. 5, I drew on Donald Schön's seminal book *The Reflective Practitioner*. Now I will build on this idea of simulation from his follow-up publication on the topic of reflection.

7.3 Educating the Reflective Practitioner in a Reflective Practicum

Donald Schön, in a presentation to the *American Educational Research Association* in 1987, proposed an environment where students can learn-by-doing. He named this a *practicum* which he described as a virtual world. This simulated virtual world represents the world of practice, but is not the world of practice (Schön, 1987). Such an environment provides a safe space where students can "act as if they were" practitioners. Furthermore, recent research concludes that since the 1980s, reflexivity has been recognised as a key professional competence by employers of graduates (Bruno & Dell'Aversana, 2018). This section will provide an overview of Schön's concept of a *reflective practicum* drawing mainly from his second major work on reflection, namely, *Educating the Reflective Practitioner* (Schön, 1990).

Schön suggests that professional education needs to be redesigned to combine the teaching of applied science, with coaching in the artistry of reflection-in-action. To achieve this, he proposes that professional schools, largely university based, should learn from other traditions of education for practice such as studios of art and design, conservatories of music and dance, athletics coaching and apprenticeship in the crafts (Schön, 1990 p. xii). In these educational traditions, people learn how to design, perform and produce, by engaging in design performance and production (ibid. p.16). Using a landscape analogy, Schön differentiates between the high ground of technical rationality, where problems are "solved" using research-based theory and techniques, with the swampy lowlands of messy and confusing problems that defy neat technical solutions. Such thinking, i.e. that academic research and its associated pedagogy prepares students for real-world practice, is increasingly being called into question. For example, he asks: In an engineering curriculum built mainly on engineering science, how should students learn how to design? Furthermore, drawing on the work of Dewey, he argues that a student in this type of environment cannot be taught what to do, but can be coached and guided on what course of action to take.

In order to enable a student learning a practice, Schön proposes the concept of a *practicum*, a purposefully designed setting in which the practice can be simulated. In the *reflective practicum,* which approximates a real-world situation, the student is facilitated to learn by doing with the supportive scaffolding of a teacher taking on the role of a coach and through interaction with fellow students. In this simulated environment, a learner comes to "think like a" professional such as an engineer or a manager. Moreover, the virtual world is constructed to represent the real world of practice where students can experiment and try things out in a low-risk "safe" setting. The practicum provides a bridge between university and practice in professional education. He gives the example of an architectural studio, where design projects are organised and managed closely patterned on what happens in an actual practice. Schön presents the logical paradox of *Meno*, described vividly by Plato, as an analogy of the "feelings of mystery, confusion, frustration and futility" experienced by students during the initial stage of learning to design (ibid. p. 83). Furthermore, like Socrates in the *Meno*, important aspects of design cannot be taught by must be "discovered and appropriated for oneself" (ibid. p. 92). Concerning the predicament of learning to design, Schön states (ibid. p. 93):

> The paradox of learning a new competence is this: that a student cannot at first understand what he needs to learn, can learn it only by educating himself, and can educate himself only by beginning to do what he does not yet understand.

The role of the teachers is now that of a coach, who in this reflective practicum (e.g. a studio) weaves "showing and telling" with the students "listening and imitating" in a spirit of mutual dialogue. When this dialogue works well, it results in reciprocal reflection-in-action enhancing the work of both the student and the teacher-coach. Consequently, a "design-like practice cannot be conveyed to students wholly or mainly by classroom teaching" (ibid. p. 162). However, Schön warns that there is a risk that the reflective practicum can "become a precious island cut-off both from the world of practice to which it refers and the world of academic courses in which it resides". The antidote he recommends is cultivation of engagement with competent practitioners and being cognisant of John Dewey's taxonomy of learning consisting of these three factors: knowledge, skill and character (ibid. p. 312). Finally, Schön suggest that the creation of a reflective practicum might be very advantageous to professionals in the midrange of their career as a form of continuing education, when a person has a significant amount of practice-based experience to reflect on. This would support one of the aims of this book: to provide a process that would assist the continuing professional development (CPD) of practitioners.

7.4 Generic Simulation-Action Learning (SAL) Process

Based on the case studies in Chap. 6 and the principles outlined in Chaps. 2, 3, 4, and 5, a generic simulation-action learning (SAL) process is presented in Fig. 7.1, and each phase is then explained in detail below. I am conscious that the SAL process is a distillation of my experience of working and teaching in the area of design and innovation. In Chap. 2, I presented an argument by Thomas (2011) that the case study is concerned with phronesis and that it is a model based an individual's life-world. It is personal, and it helps us make sense of situations providing meanings which are "malleable and interpretable in the context of varieties of experience". Every teacher and practitioner interested in using SAL will bring their own experiences and circumstances to a project. This section should be approached in the spirit of Gregory's argument that there are patterns in the design method that are applicable across a wide range of conditions from a physical artefact to Dante's Commedia (Gregory, 1966). My objective is to provide such a pattern, distilled from my experiences, to assist others shaping and developing an approach to their own unique practicum. One question is bound to arise: How did I engage in a process of reflection concerning my work? My response is that this chapter is a fruit of such reflection, honed over several years developing and implementing the process. I owe a debt of gratitude to the many students, practitioners and colleagues who facilitated this ongoing aspect my learning.

In Chap. 6, I presented earlier developments of the process which used the term "step" for each section of the project's progression. I have changed the terminology to phase, based on feedback and further reflection, in order to avoid giving the impression that the process involves stepping linearly through a rigid process.

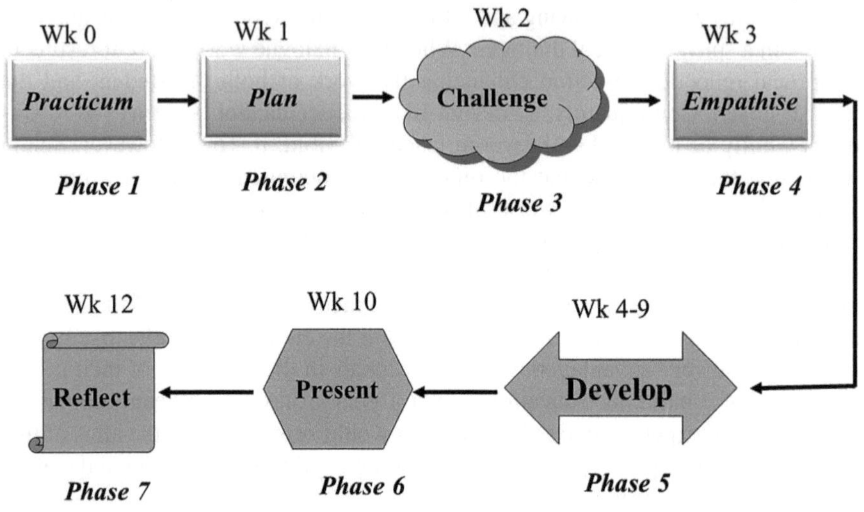

Fig. 7.1 SAL module roadmap (linear)

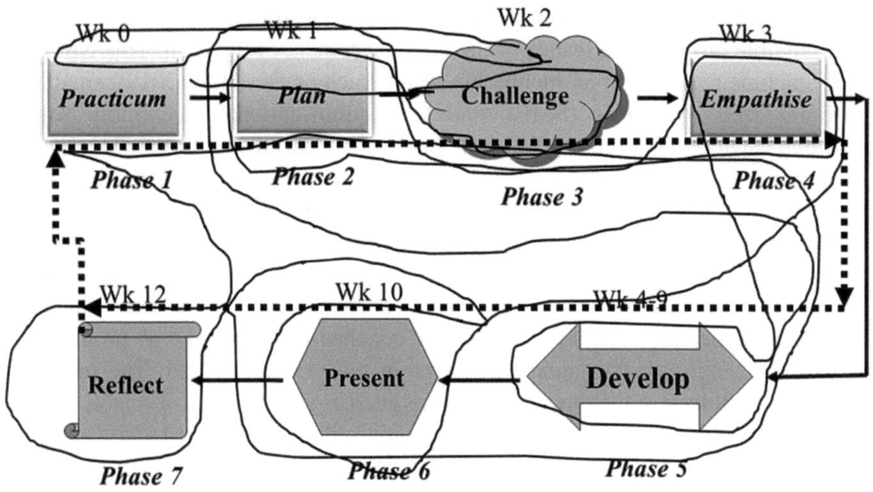

Fig. 7.2 SAL Module Roadmap (reality)

The aim is to provide teachers with a phased process to implement simulation-action learning in an academic environment but also for providers of continuing professional development. The week numbers are indicative only and can vary according to the needs of the project. While the phases are showing in a serial fashion, the reality is that there will be overlaps and iterations particularly in the development phase. I have tried to capture the reality of a design journey in Fig. 7.2. The figure is rather messy, and this is the intention. The reflection process encompasses all the phases, and as a result, it is shown as a continuous circling dotted line. Recall Donald Schön's words quoted in Chap. 5 that challenging problems exist in the "swampy lowland" of messy situations that do not respond to neat technical solutions.

Note: Fig. 7.1 assumes a 14-week semester. However, year-long projects can be accommodated by adjusting the week numbers accordingly.

Furthermore, I will list examples of types of learning that can be built into each phase using Laurillard's taxonomy of conventional and digital technologies presented in Table 1.1 of the introduction chapter. It is important to be conscious that teachers are also proto-learners in the SAL process. Now I will proceed through each phase outlining the purpose of the phase, the content of the phase and expected learning from the phase.

7.4.1 Phase 1: Practicum

Purpose *To set up the reflective practicum (or virtual workplace) comparable to that described by Donald Schön above*

The module teacher will generally contact an industry/service partner to establish possible projects in advance of the commencement of the term. This will typically include start-ups in an incubation centre such as the iHub (2019), small and medium-sized companies (SME), and large multi-nationals preferably with a research capability. Partners can also include not-for-profit enterprises. The specifications for eternal competitions, such as the Enterprise Ireland Student Awards described in Chap. 6, can supplement (or assume) the role of a client. Where possible, cross-functional teams should be set up at this stage with preferably three to five members. This will involve the teacher collaborating with other departments to establish a cooperation with a like-minded teacher committed to the idea of interdisciplinary partnership. The students will be coached to "think like a" professional practitioner, e.g. engineer and/or manager in the "virtual" world. The reflective practicum will have some physical locations such as the client's workplace, student laboratory or tutorial rooms. However, the teams will be encouraged to use any suitable spaces for meeting and sharing ideas and plans.

In an ideal situation, the collaborating teachers share their timetable and location for a weekly lab/tutorial. Moreover, this would provide an opportunity for the lecturers to team-teach and share their professional and academic experience with the student teams where they assume the roles of mentors and coaches.

My reflection on this initial phase was greatly influenced by my encounter with Aristotle's teleological worldview as described in Chap. 2. The origin of the concept is in the Greek word "telos" which the Oxford Dictionary explains as an *ultimate object or aim*. It challenged me to set out my aim for my students. Consequently, I expressed to them that my overall objective was that the SAL process would enable them to be *professional*, *innovative* and *trustworthy* people. Every phase, deliverable and action would be implemented with this teleological principle in mind (Bogen, 2005).

Furthermore, the metatheory developed in Chap. 4 influenced my design of the practicum. The organisation that was chosen had a track record of being innovation-based. The collaboration of the teams was an interactive process. The design and innovation challenge was present within the organisational ecology of the company, the region and also of the educational institute.

Types of Learning The main learning type in phase one involves ***acquisition*** by the teachers of a suitable client through face-to-face meetings with, for example, incubation centre managers and other potential industry/services practitioners. The teachers can be informed through regular ongoing reading of industry newsletters, relevant websites and external seminars associated with the themes of design and innovation. The initial gathering of ideas and preparation for SAL can start even months before the project begins. SAL usually involves more pre-semester preparatory work than conventional lecturing.

7.4.2 Phase 2: Plan

Purpose *To agree a plan for the semester (or year)-long collaborative project*

The teacher(s) meet with the client to further explain the pedagogical approach and to clarify requirements and deliverables. This is an important phase in developing a relationship with the practitioner at the beginning of the extended interaction. However, it is worth stressing that work for the practitioner should be kept at a very reasonable level given their busy workload. Any questions or clarification can be channelled through the teachers to prevent unnecessary duplication and unclear questions. It is recommended that each student team is tasked with developing a project plan (e.g. using MS project) showing the estimated milestones and deliverables. This initial planning exercise can be unnerving for the students who ask: how can we plan a project that is not yet specified? However, the timeline of the semester and competition deadlines will indicate the timescales of the main project milestones. For example, the design thinking process that was shown in Chap. 1 will provide headings for the plan that can be modified later. Furthermore, the plan will be a live document that will be updated weekly by the students, based on how the project unfolds, and as deliverables are clarified and even rescheduled. This is an aspect that must be recognised given the contingencies and unforeseeable circumstances associated with real-life projects.

Types of Learning Here students will carry out ***practice*** exercises using project planning simulation and software. Reference to examples of project plans for historical design and innovation projects will assist practice in this phase. The students are encouraged to complete an exercise of developing an industry standard template for team meeting minutes that includes inter alia: agenda, attendees, action lists complete with responsibilities and timelines.

7.4.3 Phase 3: Challenge

Purpose *To specify the design problem that will challenge the students to develop and present a solution by the end of the semester*

The practitioner completes a short description of the design problem and sends it to the teacher(s) to review. This draft design brief can then made available to the students using a virtual learning environment (VLE) such as Moodle. The teacher(s) meet with the class and presents an overview of the module learning outcomes and the structure of the project as well as assessment criteria and expected project logistics. Then the class is divided into project teams that review the draft design problem and prepare for a meeting with the practitioner. Michaelsen's (2004) principles of team-based learning outlined in Table 1.2 can be utilised in this phase. The teams will develop a written summary of their interpretation of the design project proposal with a list of associated questions in preparation for the face-to-face meeting in phase four. Here the ground rules of the project are reiterated to the students, i.e. they assume responsibility for all decisions and must from now on "think like a" professional engineer or manager.

Types of Learning Students teams to focus on learning through ***inquiry*** by using online and print media to gain as much information about the client's organisation and their operations as possible. Such inquiry will ensure that a list of coherent and relevant questions is prepared. Here the teachers should immediately assume the role of coach by probing the research and reasoning involved in the development of the queries.

7.4.4 Phase 4: Empathise

Purpose *To clarify the design brief and "put yourself in the client's shoes" during a face-to face meeting*

The class project teams meet the practitioner (who assumes the role of a real-world client) face to face. The client presents the design problem to the class verbally with more detailed description than in the design brief. The preferable location of this meeting is in the client's workspace and ideally involves a tour of their facility. This provides an opportunity for the class to get a more in-depth view of the clients' thinking and to put themselves in the client's shoes (Leonard & Rayport, 1997). Also, the project teams have the opportunity to question the client based on their initial week-long research into the problem domain. It is recommended that a date is set when each project team will present their design solution to the client at the end of the semester. This date will be essential to planning times of project deliverables. Also issues like intellectual property (IP) are discussed at this point, and in some cases the students and teachers will be asked to sign a non-disclosure agreement (NDA).

Types of Learning This meeting is crucial to the students' and teachers' deep learning about the client's workplace and business through ***acquisition*** and through ***inquiry*** in this face-to-face context.

7.4.5 Phase 5: Develop

Aim *To develop a solution to the design problem using a team-based approach over the course of the semester (or year)*

Each week the project teams present a status of their work to the lecturer who in this type of pedagogy acts as a coach and advisor rather than the conventional lecturing mode. The project teams work on the design problem during the semester using academic and industry standard design and innovation methodologies (Cooper, 2001; Cross, 2011; Curedale, 2013; Ulrich & Eppinger, 2004). Providing a critique or an overview of which methodology to use is outside the scope of this book. However it is recommended that the teacher becomes familiar with at least one process and provides an overview to the students (Costello, 2015). The project teams update their detailed project plan in the form of a Gantt chart subsequent to the client meeting and review/revise the plan each week. The project report can be started immediately with an overview of the client's organisation presented to the teacher(s). Such presentations are normally informal in nature and involve discussions with each team individually. Some aspects of "agile" development methodologies are recommended (agilemanifesto, 2001; Christopher, 1998; Conboy, 2009). For example, each class/tutorial should begin with a short and focused stand-up meeting (Pikkarainen, Wang, & Conboy, 2007). Each team will briefly answer the following three questions: What was completed in the last week? What are the plans for next week? Are there any issues that needs to be resolved that requires input from the client or other stakeholders? Such queries should be channelled through the teacher-coach. The agile literature refers to these short interval meetings and worktime periods as "sprints". I have divided these sprints into two sections when working with the student teams: "The Project Scrum", and what I call "The Reflective Scrum". The Reflective Scrum also involves three questions, in the same way as the Project Scrum: What have you have learned since the last phase (about the project and about yourself)? What do you need to learn for the next phase? Are there any impediments to enabling this learning? In practice, I find some students somewhat reticent about publicly providing a reflection on what they have learned about themselves. However, I believe that the constant reference to the importance and frequent practice of reflection keeps this attribute constantly in their attention. Furthermore, the routines are designed, using Aristotle's concept of "habit" described in Chap. 2, so that reflection-in-action becomes imbedded in the student's awareness. During the Develop phase, I ask the students to provide a one page "poster" that visually describes their concepts and I send these to the practitioner. This interim review provides the practitioners with an idea of how the teams are progressing and gives them the opportunity to provide feedback to the teams. This feedback is very beneficial to the students who can improve their proposed solutions before the final presentation to industry practitioners.

Laurillard emphasises the importance of dialogue for learning, and the SAL process facilitates this interaction between teacher and student in a team-based learning environment. SAL is not modelled on the traditional lecture "transmission" model and requires a different mind-set and resource allocation to implement successfully. Figure 7.1 shows the flowchart for this phase as a symbol consisting of arrows pointing in both directions. The idea here is to convey that the "Develop" phase is a non-linear multifaceted process, with many iterations and loops that can go back and forth.

Types of Learning The group project will enable learning through ***collaboration***, and it was seen in the case studies in Chap. 6 that collaborating in interdisciplinary teams broadened the students' learning horizon. Also, the students can set up online fora for cooperating outside of class times. A natural output from the collaboration is the development of ***discussion*** opportunities in groups or one to one. Many Internet tools are available to continue such dialogue within the teams and with the teacher-coaches. Team teaching by lecturers from different disciplines (e.g. engineering and business) can provide the teams with different perspectives on resolving problems and developing solutions. Using the real-world project scenario, having to "think like a" practitioner and using the project world as a virtual "lab" give an ample opportunity for learning through ***practice***.

7.4.6 Phase 6: Present

Purpose *To present the design solutions to internal and external stakeholders*

The class project teams present their design solutions to the practitioner (or a team from the client's organisation) and teacher(s) through oral presentation and a project report. It is advisable that the date for this oral presentation is fixed early in the project. Where the project involves a student enterprise competition, it is valuable to have the student teams present their ideas and prototypes to the incubation centre managers. Such business-focused feedback is an important learning experience for the students. The practitioner may like to present an award for the best design and also for the best presentation. However, it is essential that the practitioner provides feedback to the student teams in general terms and specifically in relation to their future professional careers. The prizes can be minimal (e.g. company marketing paraphernalia), but it adds to sense of occasion. This latter point may seem trivial, but, in my experience, having students present to external professionals results in "upping their game", and the presentation "event" become an important milestone and showcase. The teacher-coach can also solicit feedback from the practitioner on suggestions for improving the SAL process itself and the mode of interaction with an external organisation. The project deliverables include such items as a set of working drawings, computer-aided design (CAD) models and/or renderings, a business case. An artefact such as a mock-up of the design in cardboard or other materials is encouraged but not mandatory. This early development of an artefact is now sometimes called *preto-typing* in the literature, and it is also referred to as "fake it before you make it" (Pretotyping, 2019). Incubation centre managers are impressed when a business idea is presented in a tangible form, however rudimentary it is. A video can be included in the suite of presentations as this medium is increasingly being used both in industry and service sectors to demonstrate concepts and solutions. This is a core requirement for the Enterprise Ireland Student Awards competition (Enterprise Ireland, 2019).

Types of Learning This phase will involve the ***production*** of many artefacts such as models (physical and virtual), videos, slideshow presentations, design reports and business cases. The students should be encouraged to design innovative means of presentation and communication. Storyboarding and development of personas are powerful means of achieving this type of learning.

7.4.7 Phase 7: Reflect

Purpose *To engage in a reflection on the project using a structured approach such as the one developed in Chap. 5*

Reflection and feedback from the students are an ongoing feature of the SAL process but are also formally built into the module review process at the end of the semester. The Develop phase above described the Scrum meetings and the ongoing growth of a reflective habit in the student's learning. There is a scarcity of pedagogical tools that support the development of reflection-in-action, and I hope that my proposal of "The Reflective Scrum" can contribute something to this debate. At the end of the module, each student is required to do an assessment of their own contribution to the project. Teachers might consider a number of communication channels, such as written journals, oral presentations and face-to-face interviews. The rationale used for this is based on the author's experience of having to complete reviews in industry. The feedback is two-way, which is important for the lecturer who is continually endeavouring to improve the module content and the pedagogical process year on year. Projects can be assessed, and the same mark given to all students in a project team for presentation and reports. However, the end of project report and the one-to-one interview can be individually marked with a significant overall module weighting. Typical assessment criteria include innovation, technical content, feasibility, teamwork, construction of an artefact and presentation skills. The individual interview helps the students to orally present and defend their work. Furthermore, this simulates the situation of undergoing an interview for a future job.

Types of Learning Chapter 5 proposed that the process of reflection turns experience into learning and that it involves serious thought and consideration during each phase of the teaching process. The aim of the chapter was to provide guidance for students, teachers and practitioners on the process of reflection. This is in the context of a paucity of guidance on how to, in practice, undertake "reflection". Table 5.1 sets out a series of questions to aid the process: informally during a project and formally towards the end of a project. Furthermore, Laurillard's conversational framework places reflection as a key aspect of the learning process (Laurillard, 2013).

7.5 A SAL Process for CPD

This section will describe how the SAL process can be adapted for continuing professional development (CPD) that will involve the education and training of practitioners rather than academic students. It will suggest how each phase can be modified to suit practitioners' circumstances and timeframes.

Phase 1 – Practicum: In this application of the process, the reflective practicum can be an actual project undertaken in the organisation by the team members.

However, there is a need for a suitable (internal or external) coach to enable the process. Where several people from different organisations wish to undertake SAL, it could be condensed into a one-day or two-day intensive project with multidisciplinary teams addressing a specific organisational problem. A word of caution: when multiple organisations work together on a project, there can be intellectual property (IP) issues to deal with. In this case it may be prudent to set the teams to work on a neutral so-called wicked problem, for example, in the areas of the environment, healthcare or education.

Phase 2 – Plan: The approach to planning will essentially be the same as in the SAL process above. However, a one-day or two-day programme will require a short-term detailed project plan.

Phase 3 – Challenge: The challenge in this case can be presented by someone independent of the teams and who is responsible for management of the design problem. It is recommended that the person should be familiar with the idea of role-playing and the SAL rational and be autonomous from the team members.

Phase 4 – Empathise: Teams members will need to gather information from people who are stakeholders in the design problem through interviews or focus groups. In short interval courses, the organiser may require people to act in this role and will need to appoint them in advance.

Phase 5 –Develop: Teams will be required to work intensively to develop a solution to the design problem. Organisers will appoint suitable coaches who are familiar with the SAL process aims and objectives. Operations such as prototyping will use basic materials such as cardboard, tape and paper.

Phase 6 – Present: The time, duration and location of the presentation will be provided by the organiser at the start of the CPD event. The media available for the presentation, e.g. slideshow, video and oral performance, will also be specified at this time of the project.

Phase 7 – Reflect: The approach to reflection will be like the generic SAL process above with the frequent use of Scrum sprints instilling the habit of reflection. Such group reflection is to be recommended. In addition, it may be easier to accommodate in the CPD situation where no academic marks and grades are involved.

7.5.1 Challenges with the SAL Framework

Engaging with many cross-functional teams is a challenge for teaching, especially in the context of assuming the role of teacher as coach. However, in a team-teaching context, the teams can be limited to a relatively short interaction with each teacher based on processes such as Scrum (2019). Once given direction, the teams can work together in any suitable location but must regularly demonstrate to the teachers that progress is being made on the project and that deliverables are being met. The thrust of a SAL project is deliverable-driven rather than class time-driven. This can be disconcerting for all involved, but modern organisations place much more emphasis on *what is achieved* over *the time spent* in a fixed location. The setting-up of the practicum needs to be crafted to the circumstances of the educational institution, the experience of the teacher-coach and the relationship with outside organisations such as multi-national companies or incubation start-ups. The development of a practicum has much in common with the challenges and uncertainties characteristic of a real-world design and innovation project. However, despite the challenges, I personally agree with Fink (2004 p. 25) that "it transforms the joy of teaching!" I get great satisfaction observing my students progress from an initial stage of bemusement to producing a solution to a design problem that is recognised positively by practitioners and academics. Witnessing their sense of achievement is a great incentive to continue refining the approach, and it also provided a motivation for writing this book. However, it would be unrealistic to claim that there are not problems within teams in the course of simulation-action learning. I have occasionally experienced glitches such as social loafing, free riding and occasions where one member of a team has a much greater expectation and ambition for high grades than others. I have used two strategies in dealing with this aspect of a "practicum" that mirrors real life with its positives and problems. First, I refer to Michaelsen (2004) since he discusses these situations in his chapter on team-based learning and gives good direction on pre-empting and dealing with difficulties. Second, I try to use my own "phronesis" based on my experience in teaching and dealing with students. Sometimes you just have to read a particular situation, listen to all sides and steer a way forward. Developing resilience and strategies to overcome issues and problems is also a very relevant learning experiences for students.

7.6 Conclusions

The current literature largely associates simulation-based learning (SBL) with computer applications such as games. However, this chapter disagrees with a purely technological interpretation of "simulation" widespread in the design and innovation literature. To address this situation, the concept of simulation-action learning (SAL), developed from the empirical evidence presented in Chap. 6, is proposed as an enhancement of problem-based learning. In order to facilitate this type of learning, the chapter adopts Donald Schön concept of a *practicum*, a purposefully designed setting in which practice can be simulated. The *reflective practicum* is designed to approximate a real-world scenario, utilises available resources, includes interactions with practitioners and emerges from the imagination of the teacher whole develops a virtual world. Here the student is facilitated to learn by doing within the supporting structure of a teacher taking on the role of a coach and through interaction with fellow students. In this simulated environment, a learner is coached and challenged to "think like a" professional such as an engineer or a business manager. In addition, the SAL process can be adapted to workplace education, and training needs to support practitioner continuing professional development (CPD) needs.

Kurt Lewin is famous for his assertion that there is nothing as practical as a good theory. However, his student Urie Bronfenbrenner (2005, p. 48) reversed this classical Lewinian maxim to read: there is nothing like the practical to build a good theory (p 48). This chapter seeks to contribute to the debate in contemporary technological education, by building a schema that binds together the world experiences, described in Chap. 6, into a simulation-action learning (SAL) framework.

Chapter Summary

- There has been an increasing interest in the concept of "simulation" in the academic literature.
- The application of simulation to learning has largely been associated with computer-based environments.
- The concept of simulation is currently used in areas such as medicine to approximate real-world scenarios.
- Donald Schön's book *Educating the Reflective Practitioner* advocates the notion of a *reflective practicum* to approximate a real-world scenario.
- These ideas and the practice-based cases outlined in a previous chapter have resulted in the development a theoretical framework: simulation-action learning (SAL).
- The SAL phases are presented in diagrammatic form and outlined in detail to assist implementation in an educational context.
- SAL is presented as phases rather than steps to indicate the iterative and back-and-forth nature of the design process.
- SAL can also be adapted for continuing professional development (CPD) courses.

References

agilemanifesto. (2001). Manifesto for Agile Software Development available on line through http://agilemanifesto.org/. Accessed July 2019.

Bogen, J. (2005). Teleological explanation. In T. Honderich (Ed.), *The Oxford companion to philosophy* (2nd ed., p. 911). Oxford: Oxford University Press.

Bronfenbrenner, U. (2005). Lewinian space and ecological substance. In U. Bronfenbrenner (Ed.), *Making human beings human: Bioecological perspectives on human development* (pp. 41–49). Thousand Oaks, California: Sage Publications.

Bruno, A., & Dell'Aversana, G. (2018). Reflective practicum in higher education: The influence of the learning environment on the quality of learning. *Assessment & Evaluation in Higher Education, 43*(3), 345–358. https://doi.org/10.1080/02602938.02602017.01344823

Cadotte, E. (2014). The use of simulations in entrepreneurship education: opportunities, challenges and outcomes. In M. H. Morris (Ed.), *Annals of entrepreneurship education and pedagogy -2014* (pp. 280–304). Cheltenham Edward Elgar, Publishing Limited.

Cant, R. P., & Cooper, S. J. (2010). Simulation-based learning in nurse education: Systematic review. *Journal of Advanced Nursing, 66*(1), 3–15.

Casar, J. R. (2000). Encouraging students' attitude of innovation in research universities. *European Journal of Engineering Education, 25*(2), 115–121.

Christopher, M. (1998). Creating agile supply chains. In J. Gattorna (Ed.), *Gower handbook of supply chain management*. Aldershot, UK/Gower, MO: Burlington.

CISL. (2019). Center for immersive and simulation-based learning avialable on line through http://cisl.stanford.edu/. Accessed July 2019.

Conboy, K. (2009). Agility from first principles: Reconstructing the concept of agility in information systems development. *Information Systems Research, 20*(3), 329–354.

Cooney, T. M., & Murray, T. M. (2008). Entrepreneurship education in the third-level sector in Ireland avialable on-line through http://arrow.dit.ie/imerep/2/. Accessed Dec 2019: Institute of Minority Entrepreneurship, Dublin Institute of Technology.

Cooper, R. G. (2001). *Winning at new products : Accelerating the process from idea to launch*. New York: Perseus Publishing.

Costello, G. J. (2015). Working paper: Review of innovation process frameworks *Research@ THEA Repository https://research.thea.ie/handle/20.500.12065/361*. Accessed July 2019.

Creed, C. J., Suuberg, E. M., & Crawford, G. P. (2002). Engineering entrepreneurship: An example of a paradigm shift in engineering education. *Journal of Engineering Education, 91*(2), 185–195.

Cross, N. (2011). *Design thinking: Understanding how designers think and work*. Oxford: Berg Publishers.

Curedale, R. (2013). *Design thinking : Process and methods manual*. Design Community College Inc. Topanga, California

Davis, M. (2017). *Teaching design: A guide to curriculum and pedagogy for college design faculty and teachers who use design in their classrooms*. New York: Allworth.

Deegan, M., Stave, K., MacDonald, R., Andersen, D., Ku, M., & Rich, E. (2014). Simulation-based learning environments to teach complexity: The missing link in teaching sustainable public management. *Systems, 2*(2), 217–236.

Eck, R. V. (2006). Digital game-based learning: It's not just the digital natives who are restless.... *Educause Review, 41*(2), 16–30.

Enterprise Ireland. (2019). Enterprise Ireland Website http://www.enterprise-ireland.com/en/. Accessed Dec 2019.

Fadale, K., Tucker, D., Dungan, J., & Sabol, V. (2014). Improving nurses' vasopressor titration skills and self-efficacy via simulation-based learning. *Clinical Simulation in Nursing, 10*(6), 291–299.

Fink, L. D. (2004). Beyond small groups: Harnessing the extrordinary power of learning teams. In L. K. Michaelsen, A. B. Knight, & L. D. Fink (Eds.), *Team-based learning: A transformative use of small groups in college teaching*. Sterling, VA: Stylus Publishing.

Gregory, S. A. (1966). *The design method*. Springer US. https://doi.org/10.1007/978-1-4899-6331-4

iHub. (2019). GMIT innovation Hubs https://www.gmit.ie/gmit-innovation-hubs. Accessed July 2019.

Jørgensen, F. (2004). The student voice. In A. Kolmos, F. K. Fink, & L. Krogh (Eds.), *The Aalborg PBL model: Progress, diversity and challenges* (pp. 21–35). Aalborg, Denmark: Aalborg University Press.

Khunger, N., & Kathuria, S. (2016). Mastering surgical skills through simulation-based learning: Practice makes one perfect. *Journal of Cutaneous and Aesthetic Surgery, 9*(1), 27–31.

Kolmos, A., Fink, F. K., & Krogh, L. (2004). The Aalborg model - problem based and project-organized learning. In A. Kolmos, F. K. Fink, & L. Krogh (Eds.), *The Aalborg PBL model: Progress, diversity and challenges* (pp. 9–19). Aalborg, Denmark: Aalborg University Press.

Krogh, L., & Rasmussen, J. G. (2004). Employability and problem-based learning in project-organized settings at universities. In A. Kolmos, F. K. Fink, & L. Krogh (Eds.), *The Aalborg PBL model: Progress, diversity and challenges*. Aalborg, Denmark: Aalborg University Press.

Lappalainen, P. (2011). Development cooperation as methodology for teaching social responsibility to engineers. *European Journal of Engineering Education, 36*(6), 513–519.

Lateef, F. (2010). Simulation-based learning: Just like the real thing. *Journal of Emergencies, Trauma, and Shock, 3*(4), 348–352.

Laurillard, D. (2013). *Rethinking University teaching : A conversational framework for the effective use of learning technologies*. Retrieved from http://search.ebscohost.com/login.aspx?direct=true&db=e000xww&AN=606935&site=ehost-live. Hoboken: Routledge.

Leonard, D. A., & Rayport, J. (1997). Spark innovation through empathic design. *Harvard Business Review, 75*(6), 102–113.

Lindblom, C. E. (1959). The science of "muddling through". *Public Administration Review, 19*(2), 79–88.

Lu, J., Hallinger, P., & Showanasai, P. (2014). Simulation-based learning in management education. *Journal of Management Development, 33*(3), 218–244.

Luryi, S., Tang, W., Lifshitz, N., Wolf, G., Doboli, S., Betz, J. A., et al. (2007). *Entrepreneurship in engineering education*. Paper presented at the frontiers in education conference-global engineering: knowledge without borders, opportunities without passports, 2007. FIE'07. 37th Annual.

Mendelson, M. I. (2001). Entrepreneurship in a graduate engineering program. *Journal of Engineering Education, 90*(4), 601–607.

Michaelsen, L. K. (2004). Getting started with team-based learning. In L. K. Michaelsen, A. B. Knight, & L. D. Fink (Eds.), *Team-based learning: A transformative use of small groups in college teaching* (pp. 27–50). Sterling, VA: Stylus Publishing.

ODE. (2006). Simulate. In *Oxford dictionary of english* (Revised, 2 ed.). Oxford: Oxford University Press.

Ohland, M. W., Frillman, S. A., Zhang, G., Brawner, C. E., & Miller III, T. (2004). The effect of an entrepreneurship program on GPA and retention. *Journal of Engineering Education, 93*(4), 293–301.

Papayannakis, L., Kastelli, I., Damigos, D., & Mavrotas, G. (2008). Fostering entrepreneurship education in engineering curricula in Greece. Experience and challenges for a technical university. *European Journal of Engineering Education, 33*(2), 199–210.

Perrenet, J. C., Bouhuijs, P. A. J., & Smits, J. G. M. M. (2000). The suitability of problem-based learning for engineering education: Theory and practice. *Teaching in Higher Education, 5*, 345–358.

Pikkarainen, M., Wang, X., & Conboy, K. (2007). *Agile practices in use from an innovation assimilation perspective: a multiple case study*. Paper presented at the international conference on information systems (ICIS 2007), December 10th – 12th, Montreal, Canada.

Prensky, M. (2001). Digital natives, digital immigrants. *On the Horizon (NCB University Press), 9*(5), 1–6.

Pretotyping. (2019). Pretotyping.org https://www.pretotyping.org/. Accessed July 2019.

Ruse, M. (2005). Theory. In T. Honderich (Ed.), *The Oxford companion to philosophy* (2nd ed.). Oxford: Oxford University Press.

Sara, R. (2011). Learning from life: Exploring the potential of live projects in higher education. *Journal for Education in the Built Environment, 6*(2), 8–25.

Schön, D. A. (1987). Donald Schön's presentation "Educating the reflective practitioner" to the 1987 meeting of the American Educational Research Association Washington, DC available on-line through http://post.queensu.ca/~russellt/howteach/schon87.htm. Accessed July 2019. Retrieved from.

Schön, D. A. (1990). *Educating the reflective practitioner: Toward a new design for teaching and learning in the professions*. San Francisco: Wiley.

Scrum. (2019). Scrum Alliance: Transforming the world of work available on-line through https://www.scrumalliance.org/. Accessed Oct 2019.

Silva, A., Henriques, E., & Carvalho, A. (2009). Creativity enhancement in a product development course through entrepreneurship learning and intellectual property awareness. *European Journal of Engineering Education, 34*(1), 63–75.

Smith, K. A., Sheppard, S. D., Johnson, D. W., & Johnson, R. T. (2005). Pedagogies of engagement: Classroom-based practices. *Journal of Engineering Education, 94*(1), 87–101.

Sørensen, O. J. (2004). New realities and the implications for problem-based learning. In A. Kolmos, F. K. Fink, & L. Krogh (Eds.), *The Aalborg PBL model: Progress, diversity and challenges* (pp. 109–128). Aalborg, Denmark: Aalborg University Press.

Steadman, R. H., Coates, W. C., Huang, Y. M., Matevosian, R., Larmon, B. R., McCullough, L., et al. (2006). Simulation-based training is superior to problem-based learning for the acquisition of critical assessment and management skills. *Critical Care Medicine, 34*(1), 151–157. https://doi.org/10.1097/01.ccm.0000190619.42013.94

Tan, S. S., & Ng, C. K. F. (2006). A problem-based learning approach to entrepreneurship education. *Education + Training, 48*(6), 416–428.

Thomas, G. (2011). *How to do your case study: A guide for students and researchers*. Thousand Oaks, California: SAGE Publications.

Ulrich, K. T., & Eppinger, S. D. (2004). *Product design and development* (3rd ed.). Boston/ London: Irwin McGraw-Hill.

Chapter 8
Contemporary Issues for Design and Innovation Studies

8.1 Introduction

The aim of this chapter is to respond to the challenges set by the editors of this series: to stimulate debate and push boundaries in technology education. The sections are written in the form of vignettes for investigation and reflection by people involved in the teaching and research of design and innovation. Each vignette is followed by suggested questions for discussion and further study. The topics have been assembled from the author's research into the management, theory, philosophy and teaching of design and innovation. The list is by no means exhaustive but is designed to provide a primer for teachers and others interested in the topic. Furthermore, the chapter aims to contribute to a research agenda in order to provide an impetus for developing more detailed studies in the area. The discussion will be divided into two areas which are important contemporary issues for the teaching of design and innovation. They broadly follow the taxonomy of the book: practices and principles. Each vignette will begin with a short introduction that provides a context in relation to themes explored in previous chapters of the book. Suggestions for discussion and further study are offered after each vignette.

Practices: Contemporary Issues
- Changing business paradigms
- The digitalisation of organisations
- Innovation Technology (IvT)
- The artificial intelligence debates
- Organisational climate for design and innovation
- The innovation dilemma
- Sustainability

Principles: Contemporary Issues
- The role of virtue ethics in teaching of design and innovation
- The influence and consequences of Positivism

G. J. Costello, *The Teaching of Design and Innovation*, Contemporary Issues in Technology Education, https://doi.org/10.1007/978-3-030-41380-4_8

- Heidegger's questions concerning technology
- Teaching as a pursuit of truth
- Design, innovation and the common good
- The Wizard-of-Oz

8.2 Preamble: The Journey from the "Potter" to the "3D Printer"

First, I will conceptualise what I consider to be the 360° circular development of design and innovation from the "Potter" to the "3D Printer". My argument here is that the Potter combined kinaesthetic "design" skills and "maker" (manufacturing) acumen in a real-time creative synergy. Furthermore, these kinaesthetic abilities were the preserve of the gifted few, usually artisans. The industrial revolution ruptured the connection between "designer" and "maker" sometimes pejoratively described as *Taylorism*. Some significant attempts were made to bridge the design-manufacturing chasm such as the influential Bauhaus movement in the twentieth century (Davis, 2017). However, the recent advent of computer-aided design (CAD) and additive manufacturing (3D printing) has provided the opportunity to restore this assimilation, which is personified in the "maker movement". Furthermore, the kinaesthetic capability imbedded in CAD and the ubiquity of inexpensive 3D printers allows many more people, than the gifted few, to have the ability to both "design" and "make" in a concurrent hermeneutic. This has implications for the role of education and pedagogical approaches.

Now, I will present vignettes of what I consider to be important themes for contemporary studies in design and innovation. This chapter starts with practice-based themes and follows with philosophical questions. These are not exhaustive and have been chosen by the author based on his research and practitioner experience.

8.3 Practices: Organisational Questions for Design and Innovation

8.3.1 Changing Business Paradigms

Context *This vignette has implications for the principle of theory outlined in Chap. 3. The metatheoretical framework placed design and innovation within a nested structure of an ecological framework. Teaching design and innovation requires an awareness of the ecology: the business and other contexts within which the education is located.*

Chesbrough (2003) argues that in many industries the centralised approach to research and development (R&D), which he terms closed innovation, has become obsolete. He argues that the closed innovation paradigm must be replaced by open innovation whereby ideas and knowledge from outside the company are adopted in conjunction with internal processes. The factors influencing this change are the mobility of skilled people, the increasing presence of venture capital, the emergent high-tech start-ups and the significant role of university research. The success of companies such as Cisco and Intel has been attributed to their adoption of this new paradigm. In contrast, many believe that a decrease in Xerox's innovation capacity was due to the company's reliance on a closed R&D system. Chesbrough argues that embracing external ideas and knowledge will actually multiply the advantage of internal efforts. However, connecting external innovation to internal innovation requires a new business model. Figure 8.1 shows an innovation funnel adapted to illustrate an open innovation model. Furthermore, technology in Chesbrough's model is replaced by "design and innovation" that can take the form of products, services and capabilities. The funnel's "open" membrane allows design and innovation to be sourced from external sources and for new products or services to be channelled outwardly to new markets through spinoffs and licencing agreements.

The growing significance of the open innovation paradigm (Pullen, de Weerd-Nederhof, Groen, & Fisscher, 2012) has prompted West, Vanhaverbeke, and Chesbrough (2006) to propose a research framework with the following classifications: individual, organisational, value network, industry/sector and national institution (p. 288). In related work, Vanhaverbeke and Cloodt (2006) suggest that emerging forms of value networks must be examined at the level of different nested layers. These diverse layers span the spectrum from the individual to firms-organisations, through dyads, onto inter-organisational networks and ultimately reaching to national/regional innovation systems.

The most tangible area for innovation is the development and introduction of products and services. However, there has been a significant move to delivering solutions in recent years where customers have migrated from ordering products and services to ordering solutions. This has involved major firms (especially in the

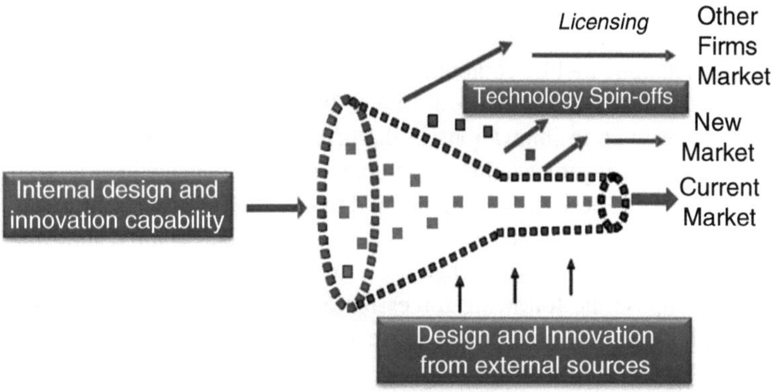

Fig. 8.1 Adaptation of Chesbrough's (2003) innovation funnel

ICT industry) positioning themselves as solution providers where the customer describes the offering they require while the company looks after the technicalities. Pricing is then done by functionality rather than hardware and software content. Another technological area that is affecting the milieu of the firm is servitisation (Dimache & Roche, 2013). Servitisation essentially describes the move on the product-service system (PSS) continuum from product plus service (e.g. product plus extended warranty) to complete service delivered through the product, the product being the enabler of services (e.g. delivery of cubic meters of cold air and not a refrigeration unit). In the current global economy, manufacturers are under pressure to continuously adapt to the ever-changing business environment (e.g. globalisation, environmental legislation, change in customer behaviour, trends towards purchasing services as distinct from products). Manufacturers can no longer rely on the traditional product-focused business models; they need to be highly adaptive and innovative in order to compete.

Suggestions for Discussion and Further Study

Provide examples of an organisation that use a "closed innovation" model and an organisation that uses an "open innovation model".

How can the concept of "open innovation" be incorporated into the reflective "practicum" of the SAL framework?

8.3.2 The Digitalisation of Organisations

Context *This vignette has implications for the notion of the "Practicum" presented in Chap. 7. The effects of digitalisation have repercussions for the situation in which organisations operate. Consequently, the teaching of design and innovation must be cognisant of this.*

The spectacular growth of the Internet, ubiquity of networking, globalisation of business and evolution of information economies has resulted in novel business processes and new ways of sharing knowledge. These transformations are resulting in the development of the fully digital firm (Laudon & Laudon, 2002). Other ICT technologies include mobile computing, teleworking, social networking, Internet of things (IoT), Industry 4.0 and "open source" that affect not only business but society. ICT has resulted in innovations in the firm affecting logistics, manufacturing, sales and order management, finance, human resource management as well as the support activities of design, engineering and marketing (O'Brien & Marakas, 2009; Post & Anderson, 2003). According to Robson (1997), "quality, innovation and service are now more important than cost, growth and control" (p 273). She also goes on to propose the following forces for openness: new technology, new geopolitical order and new enterprises. Furthermore, Robson provides the following taxonomy of the evolution of the firm:

- *Efficiency* was the price of staying in business in the 1960s.

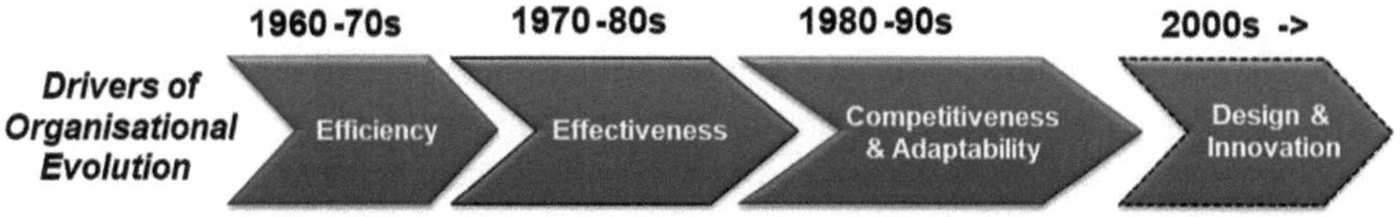

Fig. 8.2 A four-era model of organisational evolution. (Developed from Ward, Griffiths, and Whitmore (1990))

- *Effectiveness* was the price of staying in business in the 1970s.
- *Competitiveness* was the price of staying in business in the 1980s.
- *Adaptability* was the price of staying in business in the 1990s.

I argue that:

- *Design and Innovation* is the price of staying in business in the 2000s.

As Pfaffenberger (2002) puts it: "the internet has emerged as an un-paralleled public medium for communication and commerce-and it's changing our world".

Figure 8.2 diagrammatically develops this concept by building on the three-era model of Ward et al. (1990) and showing organisations entering a fourth era where the main driver is design and innovation rooted in a digital organisation.

It should be noted that this figure does not imply that design and innovation was not an important consideration for enterprises prior to the new millennium. However, ceteris paribus, it is now regarded as the differentiator in an organisation's success.

Suggestions for Discussion and Further Study

What are the implications for human-centred design resulting from the increasing digitalisation of organisations?

What are the challenges for security and privacy in the digital organisation?

8.3.3 *Innovation Technology (IvT)*

Context *This book is part of a series that explores contemporary issues in technology education. Consequently, the teacher is challenged to become familiar with the range of technologies available in the area of design and innovation. Technology has implications for how design and innovation is carried out but also for how it is taught.*

According to the economist John Kenneth Galbraith (1967), technology means the systematic application of scientific or other organised knowledge to practical tasks. The Oxford Dictionary of English defines technology as the application of scientific knowledge for practical purposes (ODE, 2006a). Technology is used "to represent things, action, processes, methods and systems" (Kline, 2002 p. 210) .

Dodgson, Gann, and Salter (2005) propose that a range of new technologies such as simulation and modelling tools, virtual reality, data mining and rapid prototyping have led to the intensification of design and innovation. They have used an umbrella term – innovation technology (IvT) – to describe these new tools and methods. IvT they argue is being increasingly applied to design and innovation and indeed is dramatically changing the nature of the innovation process. Furthermore, they contend that IvT is having a significant influence on creative tasks and the ways in which knowledge is constructed, shared and used. They describe their schema of the application of IvT to the innovation process in terms of three characteristics: thinking, playing and doing.

- *Think:* in that IvT can liberate creative people from mundane tasks and enable them to experiment more freely and widely resulting in the production of a variety of options.
- *Play:* design, prototyping and testing can be carried out more effectively and economically. Also, investment choices can be delayed until market and technology patterns become clearer.
- *Do:* the increasing "digital" integration with other types of technology provides innovators with greater confidence in their ability to transform ideas into products and services.

Furthermore, they argue that the IvT enablement of thinking, playing and doing is a major support to organisations in dealing with disruptive innovation (doing things differently) and incremental innovation (doing existing things better). In this age of avatars and the Internet of things (IoT), there is much work to be done on examining the implications of technologies such as virtual worlds and social networks for design concepts such as empathy (Potapova & Rodionov, 2014).

Suggestions for Discussion and Further Study

Suggest a suitable technology for teaching the SAL design and innovation process.

In what way will technology assist or hinder the "empathy" phase of the SAL framework?

8.3.4 Organisational Climate for Design and Innovation

Context *This section links the discussion on phronesis in Chap. 2 and the concept of simulation in Chap. 7. Replicating a design and innovation environment requires the development of a suitable climate within which the students carry out their work. A supportive or a rigid climate will impact positively or negatively on their practical judgement.*

Göran Ekvall, in his work in the University of Lund, developed an instrument for measuring organisational structure and climate for creativity and innovation (Ekvall, 1996). The instrument evolved from a Swedish research programme in the 1980s which investigated conditions that "stimulate or hamper creativity and innovation" in organisations. Significantly, following Schein (1985), he differentiated between climate and culture and argues that climate "should be regarded as a manifestation of culture". Ekvall (p.105) described climate in the following terms:

> … a conglomerate of attitudes, feelings, and behaviours which characterizes life in the organization, and exists independently of the perceptions and understandings of the members of the organization.

The model in Fig. 8.3 shows Ekvall's conceptualisation of climate as an "intervening variable" that affects the organisation's operational characteristics.

He identified ten dimensions from the theoretical literature, field research and consultancy in the area of organisational psychology. These were incorporated in a Creative Climate Questionnaire (CCQ) with questions in each of the following ten sections (Table 8.1).

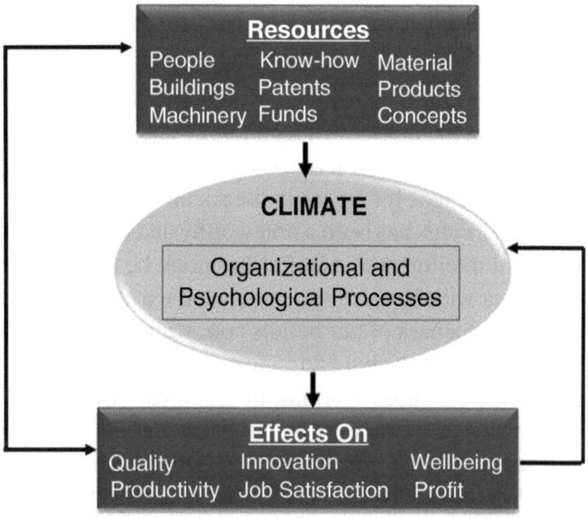

Fig. 8.3 Organisational climate as an intervening variable. (Adapted from Ekvall (1996))

Table 8.1 Eckvall's creative climate taxonomy

Challenge	The emotional involvement of the members of the organisation in its operations and goals
Freedom	The independence of behaviour exerted by the people in the organisation. For example, freedom to make contacts and give and receive information
Idea support	The ways in which new ideas are treated
Trust/openness	The emotional safety in relationships – initiatives can be taken without fear of reprisal or ridicule in case of failure
Dynamism/ liveliness	The eventfulness of life in the organisation
Playfulness/ humour	The spontaneity and ease that is displayed – relaxed atmosphere with jokes and laughter
Debates	The occurrence of encounters and clashes between viewpoints, ideas and differing experiences and knowledge
Conflicts	The presence of personal and emotional tensions (in contrast to conflicts between ideas)
Risk-taking	The tolerance of uncertainty in the organisation
Idea time	The amount of time people can use (and do use) for elaborating new ideas

Suggestions for Discussion and Further Study

How are the principles outlined in Chaps. 2, 3, 4, and 5 relevant for a design and innovation culture?

What are the implications for culture arising from "open innovation" and digitalisation of organisations?

8.3.5 The Artificial Intelligence Debate

Context *A discussion on contemporary technology and the teaching of the same invites consideration of the topic of artificial intelligence (AI). The subject raises questions for Chap. 3 which deals with the notion of responsibility.*

Herbert Simon's seminal work, *The Sciences of the Artificial* (Simon, 1996), focused on the primacy of technology and reduced the human content of design and innovation to being a more simplistic component. Thirty years later, however, the main challenges for AI are in the social sphere rather than the technological sphere. Hubert Dreyfus has been a noteworthy and controversial contributor to the debate on artificial intelligence (AI). Furthermore, he had engaged for many years in a robust argument with Herbert Simon on what we mean by the term "artificial" and what are the consequences of our beliefs concerning this topic. His work has implications for the teaching of design and innovation in the realm of AI and is a major contemporary topic for those working in this area. Dreyfus has written significantly on the whole area of technology and in particular the "mind-brain "problem. For example, his chapter in *The Philosophy of Artificial Intelligence* is relevant to this debate (H. Dreyfus & Dreyfus, 1990). Design and innovation is a fundamentally a human process and thus requires "philosophical thinking". Contemporary issues in technology education necessitate some discussion on current debates on human-technology interaction. Drawing on the work of Heidegger and Wittgenstein, Dreyfus argues that human beings are much more holistic and complex than neural nets. Furthermore he draws on the philosophical studies of Heidegger's and Merleau-Ponty to challenge many of the enthusiastic predictions from people in AI labs (H. L. Dreyfus, 2007). Consequently, the phenomenological critique of artificial intelligence is important to examine. Our conception of technology has major implications for how we teach design and innovation. Another paradigmatic shift has been the growth and diffusion of self-service technology (SST) which incorporates voice recognition, an integral technology in the quest for AI applications (Costello & Donnellan, 2007). An increasing number of business and government transactions are now being completed without human assistance. Therefore, self-service technology and business extends the traditional boundaries of the customer service function. Recently the *European Group on Ethics in Science and New Technologies* has issued an important challenge to those working in AI and related areas (EC, 2018). In their report they state that "advances in AI, robotics and so-called 'autonomous' technologies have ushered in a range of increasingly urgent and complex moral questions". The statement calls for the "launch of a process that would pave the way towards a common, internationally recognised ethical and legal framework for the design, production, use and governance" in the area.

Suggestions for Discussion and Further Study

Summarise your understanding of the mind-brain problem and its implications for AI.

Justify or refute the notion that "artificial intelligence" is really "intelligence".

8.3.6 The Innovation Dilemma

Context *This vignette draws on work from the 1970s and 1960s on innovation and proposes that it has relevance today. The authors in this piece remind us of the complexity of the design and innovation process and the risk of confusing multiplicity of ideas with the hard task of the delivery of a product or service. This would support argument in Chap. 5 that a habit of reflection is required to guard against oversimplification of the design and innovation journey.*

The work of Zaltman, Duncan, and Holbek (1973) is important when attempting to link design and innovation to organisational change. In this work, the authors define innovation as "any idea, practice or material artifact perceived as new by the relevant unit of adoption". This perception of newness moreover serves to differentiate innovation from change, and they go on to comment that while all innovations imply changes, the converse is not always the case since "not everything that an organisation adopts is perceived as new". In this work they developed their contingency theory of innovation which predicts that the effect of structural variables will be contingent on their two main sub-divisions of the innovation process: the "initiation" stage and the "implementation" stage.

Furthermore they develop, according to Slappendel (1996), "arguably the best known contingency theory of innovation". In this theory they postulate, based on their analysis of the relationship between organisational structure and innovation, that there is "no one best form for structuring organizations" (Zaltman et al., 1973, p. 134). They then identify three main characteristics of organisations which they argue facilitate both the initiation and implementation stages of the innovation-adoption process: *complexity, formalisation and centralisation.* According to their theory, the gathering and processing of information at the initiation is facilitated by higher levels of complexity and lower levels of formalisation and centralisation. However, the converse pertains during the implementation stages which are facilitated by lower levels of complexity and higher levels of formalisation and centralisation. Additionally, they introduce two "mediating factors" for each of the main stages: firstly a "high capability for effective interpersonal relations" and secondly "a high capability for dealing with conflict". Their

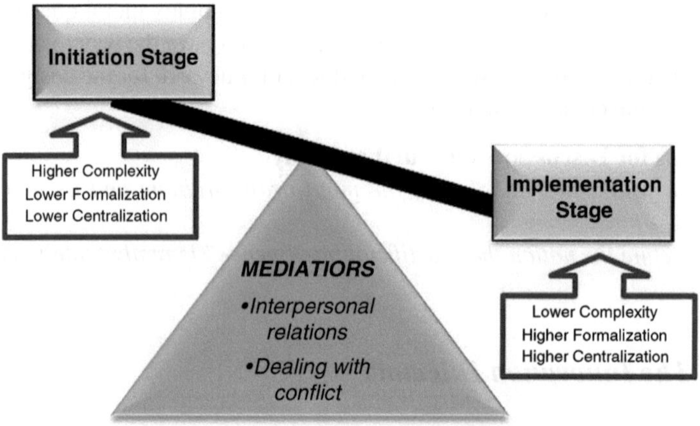

Fig. 8.4 Conceptualisation of the Zaltman et al. innovation dilemma

conclusion is that organisation must be able to adapt their structures while moving through the innovation stages. The earlier stage will require a more-organic and less-bureaucratic arrangement, while in the later stage a more mechanistic structure is called for.

According to Zaltman et al., the most important contribution by James Wilson as part of his theoretical work on innovation in the 1960s was the identification of the *innovation dilemma* which organisations face during the process of design and innovation. Wilson had concluded that it is easier to initiate than implement innovations by stating that it is "easier to increase the organizations capacity to generate new proposals than it is to increase its capacity to ratify any given proposal" Wilson (1966) cited in Zaltman et al. p. 178. Wilson had considered the characteristic of *complexity* but however did not consider *formalisation and centralisation*.

I have conceptualised the second-generation innovation dilemma proposed by Zaltman et al. in Fig. 8.4.

Suggestions for Discussion and Further Study
Describe the implications of Zaltman's model for the SAL framework.
Based on Chap. 4, evaluate whether "contingency theory" is really a "theory".

8.3.7 Sustainability

Context *Sustainability is a core value of responsibility explored in Chap. 3 and is widely regarded as a "wicked problem" facing humanity. This vignette proposes that principled design and sustainable innovation can help realise the demand by society for greater corporate social responsibility.*

Skaburskis (2008 p. 277) traces the development of the concept of a "wicked problem" to the social and technological ferment of the 1960s. In a seminar in the University of California, Berkley, West Churchman responded to a presentation by Horst Rittel, by commenting that Rittel was engaging with "wicked problems". A few years later, Rittel teamed up with Mel Webber to develop a seminal article on the topic (Kemmis & Smith, 2008). They argued that "societal problems—are inherently different from the problems that scientists and perhaps some classes of engineers deal with" (ibid. p. 160). For example, "planning problems are inherently wicked", and they went on to suggest ten characteristics of wicked problems. Pryshlakivsky and Searcy (2013) proposed the notion of sustainably as a wicked problem, while Yearworth (2016) goes so far as to say it is a "super-wicked" problem. In the context of a book on contemporary issues in technology education, it is interesting to note that Borko, Whitcomb and Liston (2008 p. 3) contend that "teaching and learning with new technologies represents a wicked problem".

The World Commission on Environment and Development was established in 1984 as a result of a UN General Assembly resolution. The report published 3 years later became known as the Brundtland Report after the name of its chairman, former Norwegian Prime Minister, Gro Harlem Brundtland. It defined sustainable development as "development that meets the needs of the present without compromising the ability of future generations to meet their own needs". The issue of sustainability and the goal of providing a sustainable quality of life is complex. Bell and Morse (Bell & Morse, 1999) contend that "though most would agree that sustainability implies *not cheating on your kids*", they consider that a clearer definition has proved to be elusive. The World Business Council for Sustainable Development (WBCSD) is an organisation made up of 200 leading international companies. They are committed to sustainable development with three objectives: economic growth, ecological balance and social progress which are also stated as eco-efficiency, innovation and corporate social responsibility (CSR). Members are drawn from more than 35 countries and 20 major industrial sectors and have committed to publicly report on their environmental performance (WBCSD, 2019). The growing interest by companies and investors in the area of sustainability is indicated by the establishment of Dow Jones Sustainability Indices which aim to create long-term shareholder value by harnessing the market for sustainable products and services (DJSI, 2019). In the area of building design, Jrade and Jalaei (2013) propose that building information modelling (BIM) technology will enable the design of more environmentally friendly and sustainable buildings. Susan Krumdieck argues for a new branch of engineering where unsustainability is investigated through the analytical treatment of "wicked problems", which sets out a programme of *how* to transition to zero carbon (Krumdieck, 2020). Over the past two decades, the fundamentals and methodologies of Transition Engineering have been developed through research and collaborations across a range of disciplines around the world.

All this is in the context of the recent international targets for sustainability, which will have major societal and business implications that will undoubtedly impact the approach to design and innovation.

Suggestions for Discussion and Further Study
How can sustainability be incorporated into the design and innovation process? Are organisations taking sustainability seriously? Discuss.

8.4 Principles: Contemporary Questions for Design and Innovation

8.4.1 Alasdair MacIntyre's "After Virtue"

Context *Chapter 2 proposed that Aristotle's concept of phronesis has relevance for contemporary design and innovation. The argument was supported by the work of several contemporary scholars in the area of social science. This vignette draws on the influential work of Alasdair MacIntyre and places the debate in a provocatively broader context.*

Alasdair MacIntyre's (2007) major work *After Virtue* first appeared almost 40 years ago and was at the forefront of a revival in virtue ethics that has continued to gather momentum (M. C. Murphy, 2003). Furthermore, Beadle and Moore (2011 p.85) have recently contended that MacIntyre "has always taken organizations seriously as objects of philosophical attention", and as a result, organisational studies can greatly benefit from his work. Kenny (2010) describes the intelligentsia of the seventeenth and eighteenth centuries as self-styled heralds, bringing the light of reason to a Europe darkened by ignorance and superstition. They contended that this dullness had resulted, to a large extent, from the religious beliefs of those whom Hume described as "the vulgar". Pointedly, Hume viewed himself as "doing for psychology what Newton had done for physics, by applying the experimental method to moral subjects" (Kenny, 2010 p. 563). Having begun in England, and been given impetus in Scotland, the Enlightenment had, by the end of the eighteenth century, become an international movement "with important social and political ramifications" (Brittan, 1999 p.266). Inwood (2005a) summarises eight leading doctrines of the enlightenment. One of these is important for this book as it views education solely as the imparting of knowledge and having no role in the development of character (which is an essential goal of virtue). According to MacIntyre (2007 p. 160), a direct consequence of the Enlightenment was the expulsion of Aristotle from European culture. It is arguable that Aristotle had been on tenuous ground since his description by Luther as "that buffoon who has misled the Church" (MacIntyre, 2002 p. 118) and the general opinion of Reformers that virtuous Christian behaviour was not essential to one's eternal salvation. Brad Gregory's comprehensive thesis that the Reformation had a direct bearing on the Enlightenment, and remains "substantively necessary to an explanation of why the Western world today is as it is", argues this point in detail (Gregory, 2012 p. 7).

The basic structure of the moral scheme that had dominated the middle ages in Europe was that which Aristotle had examined in the *Nicomachean Ethics*. Ethics, in this schema, is the teleological science that enables a man or woman transition from the state of "man-as-he-happens-to-be" to the state of "man-as-he-could-be-if-he-realised-his-essential-nature" (MacIntyre, 2007 p. 52). MacIntyre laments the spread of emotivism in a variety of philosophical guises that has resulted in it being embedded in our contemporary culture. One consequence is that the "idioms of therapy have invaded all too successfully such spheres as those of education and of religion" displacing truth as a value (ibid. p.31). One of the fundamental tenets of Aristotle's virtue ethics is the notion of teleology, i.e. that humans have a purpose and a goal. Even Kant, a giant of the Enlightenment, acknowledges that "without a teleological framework the whole project of morality becomes unintelligible" (ibid. p. 56). MacIntyre is very direct at the beginning of chapter 6 of *After Virtue*: "The problems of modern moral theory emerge clearly as the product of the failure of the Enlightenment project" (ibid. p. 62). In chapter 8 of *After Virtue*; his critique of the present-day view of managerial expertise, with its resulting consequences for organisational studies, is very relevant to the debate on design and innovation. The conventional understanding of organisational science, from the Enlightenment through luminaries such as Comte and Mill, is to provide managers with law-like generalisations to their circumstances in the same way that the laws of physics are applied to natural phenomena. In contrast, Aristotle's concept of the virtues supports a person's achievement of *eudaimonia*, variously described as happiness, blessedness and fulfilment. Virtues provide a disposition to both act and judge in particular ways to support the attainment of this *eudaimonia*. This focus on moral character is in contrast "to the approach that emphasizes duties or rules (deontology) or that emphasizes the consequences of actions (consequentialism)" (Hursthouse & Pettigrove, 2018). Decisions in organisations are summed up by MacIntyre when he says that: "choices demand judgement and the exercise of virtues require therefore a capacity to judge and to do the right thing, in the right place, at the right time, in the right way" (MacIntyre, 2007 p. 150). Consequently, judgement has an indispensable role in the life of a virtuous man which it does not or could not have in, for example, the life of merely law-abiding or rule-abiding man. MacIntyre proposes that a central virtue is that of *phronesis* and that "it comes to mean more generally someone who knows how to exercise judgement in particular cases" (ibid. 154).

MacIntyre concluded his reflection on *After Virtue* in its third edition with the following clarion call (ibid. p. xvi):

> It is also a time for resisting as prudently and courageously and justly and temperately as possible the dominant social, economic, and political order of advanced modernity.

Suggestions for Discussion and Further Study

How does a person's character influence the process of design and innovation?

Examine the implications of both virtue-ethics and deontology for design and innovation studies.

8.4.2 The Influence and Consequences of Positivism

Context *Positivism retains a highly influential position in contemporary research and teaching. This vignette is contentious and challenges the foundation of the positivist position. It has relevance to the discussion in Chap. 6 of the utilisation of qualitative "case studies" as a teaching and learning approach. It concurs with Schön's lament, in Chap. 5, regarding the effects of Positivism on contemporary education.*

The Positivist movement originated with the French philosopher and sociologist, Auguste Comte, who in the mid-nineteenth century formed a grand theory of the three stages of human thought, religious, metaphysical and scientific, with the final stage being the most productive and valuable (Lacy, 2005). "Comte believed that social reality can be explained only through science and that society's behaviour can be *determined and governed* (my italics) by natural laws. More importantly, Comte regarded all factual knowledge and phenomena a result of a predictable set of relations or a combination of relations" (Waliaula, 2013). As a result, Positivism became a significant standpoint in social research and hence its implications for the study of design and innovation. However, despite his zeal for the primacy of the scientific method based on observation and a rejection of metaphysics, Comte found it necessary to found a "religion of humanity" towards the end of his life, complete with its saints such as Frederick the Great and Adam Smith (Wernick, 2001). Comte's categories and hierarchies, of which he was rather fond, rejected Aristotle's philosophical universe where physics and metaphysics could orbit harmoniously in their different spheres (Daintith & Gjertsen, 1999). Comte's formulation of his philosophy of *Positivism* contains three principal doctrines which were outlined in Chap. 5 (Schön, 1983, p. 32). It is worth noting that according to Fotion (2005), Logical Positivism is now regarded as having run its course and has little support among professional philosophers. This is a very salient point for teachers and researchers who are still overwhelmingly positivistic in outlook.

The Vienna Circle (*Wiener Kreis*) has had a significant influence on the development of (Logical) Positivism from circa 1907 to 1930 (Bogen, 2005). The Circle flourished under the leadership of Moritz Schlick, and its membership could boast luminaries such as Neurath, Carnap and Feigl with the late but influential addition of Popper. While the *Wiener Kreis* dissolved in the late 1930s due to deaths, emigration and the rise of Nazism, its publications had a strong influence on the development of analytical philosophy (Uebel, 1999). The Circle was extreme in its antipathy to metaphysics and pursued a programme of the primacy of scientific knowledge. They saw only "one model of science for both the natural and cultural sciences" (Hirschheim, Klein and Lyytinen, 1995 p. 146).

Suggestions for Discussion and Further Study

How does the philosophical outlook of Positivism impact the process of design and innovation?

Describe a philosophical alternative to Positivism for people involved in the process of design and innovation.

8.4.3 Heidegger's Question Concerning Technology

Context *Martin Heidegger had a towering influence on twentieth-century philosophy especially the genre classified as continental philosophy. This vignette examines his view of technology and proposes that it is worth taking time to grapple with the difficult concepts, because of their influence on the realm of design and innovation. It proposes that we need to revisit Aristotle's doctrine of the mean* (Kraut, 2018) *to guard against blind technological thinking and the other extreme of developing a neo-luddite attitude.*

Heidegger in his major work *Being and Time* argued that our knowledge and basic ways of encountering the world are obtained using technology rather than by means of its scientific description. For example, a hammer is not just a tool to look at or theorise about but an implement to experience, often unconsciously, in the act of creating something. Entities in the world are not primarily objects of theoretical cognition but tools that are "ready to hand" (*zuhanden*) (Inwood, 2005b, p. 372). Furthermore Kenny (2010 p. 820) points out in his following commentary on Heidegger:

> We are not observers trying, through the medium of experience to gain knowledge of a reality from which we are detached. From the outset we are ourselves elements of the world 'always already being-in-the-world'. We are beings among other beings, acting upon and reacting to them.

Guignon (1999 p. 372) sees an "anti-humanism" in Heidegger's later works "in the description of technology (the mobilization of everything for the sole purpose of greater efficiency) as an epochal event in the 'history of being', a way things have come-into-their-own (Ereignis) rather than as a human accomplishment". Emmanuel Levinas was also critical of what he saw as Heidegger's excessive antipathy towards technology for Levinas posed the important question: Without technology "how do we feed the people"?

Macquarrie (1968 p. 52) proposes that the discussion of the significance of Heidegger's work can be organised around four topics: (1) the doctrine of man, (2) the problems of language and hermeneutics, (3) thinking and (4) the notion of being. Interestingly for disciplines such as design, innovation and technology development, Heidegger was not popular with the followers of Positivism. The Vienna Circle targeted him as a "woolly and inflated thinker whose propositions were devoid of substantial meaning", while Critchley (2001) describes what he calls the stand-off between Carnap and Heidegger.

Inwood (2005b) has the following to say about Heidegger's view of technology:

> Forgetfulness of being, 'nihilism' has culminated in the domination of the world by technology, which is not primarily machines but an event in the history of being, 'the completion of metaphysics': everything, including eventually human beings is regarded as disposable. Whether or not we can return to genuine thinking of being will determine the future of the planet.

On November 18, 1953, Heidegger presented an influential paper to the Bremen Club entitled *The Question Concerning Technology* (Heidegger, 1977). In the opening page, he emphasises that "questioning builds a way" and that we should "above all pay heed to the way". He gives a twofold answer to the question of *what it is*: technology is a means to an end and technology is a human activity. This he calls the instrumental and anthropological definition of technology. Heidegger says that the essence of modern technology is Enframing and that "Enframing belongs within the destining of revealing" (ibid. p. 13). While this notion is conceptually difficult, it is regarded as important by some scholars. For example, Bailey (2014 p. 44) argues that "Heidegger's account of Enframing (*Gestell*) is a helpful lens through which to understand the possible effects and dangers of unbridled transhumanism with respect to innovations in Human Genetic Enhancement (HGE), cloning, and human cybernetic implementation of nanotechnology to enhance cognitive and biological functions". Furthermore, the implications of the concepts of the "transhuman" and the "posthuman" for design and innovation are being explored in the literature (Forlano, 2017). The debate has even reached the more esoteric realms of educational philosophy where Dahlin (2012) argues that posthumanism challenges us to make fundamental choices about the nature of the human person.

However, Heidegger remains very controversial due to his enthusiastic support of National Socialism and his treatment of Edmund Husserl, the father of phenomenology, during the last period of his life. Some have proposed that his political views have nothing to do with his philosophy, while others consider that it places a dark cloud over his work. Due to the many facets of his life and philosophy, Moran (2000) describes Heidegger as an enigma.

Suggestions for Discussion and Further Study

Do you agree with Heidegger's view of technology or Levinas' view? Justify your position.

Describe some implication of transhumanism for the design and innovation process.

8.4.4 The Question of Truth

> Plato is my friend but truth is a greater friend – Aristotle quoted in (Kenny, 2010).

Context *This short vignette deals with a very long dispute: truth as opposed to relativism. This is pertinent to the "telos" of design and innovation explored in*

Chap. 2. It has implications for the principle of responsibility in Chap. 3, especially if the ethical output of design and innovation is considered a matter of opinion.

The question to be addressed here is the concern about a drift towards relativism, and even nihilism (Weber, 2003a, 2003b, 2004) captured in the following question: Does truth exists and is it possible to attain? This is, according to Doctor Johnson, of fundamental importance for "without truth, there must be dissolution of society". Maritain (1932) describes how Socrates confronted the mercenary tendencies of the Sophists by reforming philosophic reasoning and directing it to seek nothing but the truth and supports the school of Aristotle-St. Thomas Aquinas that truth is neither impossible or easy but is difficult to attain. He also argues that the denial of truth can be confronted by a *reductio ad absurdum* using the following line of argument: when someone says *that they do not know whether any proposition is true*, then either:

- they do in fact know this proposition is true; in which case they contradict themselves OR
- they do not know whether it is true; in which case they are condemned to absolute silence –even mental silence (Maritain, 1932, p. 181).

Scruton (2004) describes relativism as telling us "that there is no such thing as objective truth, since all truth is relative". I propose that this debate is pertinent to the teaching of design and innovation given the influence of social constructivism which some scholars find to be "deeply problematic" (Slife & Richardson, 2011).

Suggestions for Discussion and Further Study
Explain the term "reductio ad absurdum" and its implications for academic debate. How does a post-truth philosophy have implication for design and innovation?

8.4.5 Design, Innovation and the Common Good

Context *This vignette is linked to the concept of "the good" in the writings of Aristotle explored in Chap. 2 and Emmanuel Levinas' concept of the other outlined in Chap. 3. Furthermore, it challenges a tendency to extreme individualism in contemporary design and innovation projects and programmes.*

According to Murphy (1999), the common good is a "normative standard in Thomist and neo-Thomist ethics for evaluation of social, legal and political arrangements" and stems from Aristotle's understanding of humans as social and political beings. Furthermore, he contends that "any good that is a genuine perfection of our common human nature is a common good" (ibid. p 161). Jacques Maritain was one of the most prominent neo-Thomists of the twentieth century (Sweet, 2004). He defended the system of Aristotle and St. Thomas Aquinas (Maritain, 1932) in various philosophical branches. Maritain's scheme proposed that the ontology of "substance and accident" of Aristotle-St. Thomas safeguarded the uniqueness of every human being. This was contrasted with the "substance-only" construction of Spinoza, built on a Cartesian foundation, which resulted in the conclusion that everything that exists is formed of a single material. The other extreme lead by Kant denied substance and considered accidents as the only reality resulting in a concept of phenomena which Maritain regarded as chimerical and figments of imagination. Maritain considered the human being as an integrated whole with material and spiritual dimensions participating in a society ordered for the common good (Maritain, 2002). He presented the Thomist noetic that he adhered to as follows:

> It is a noetic the recognises the existence of things outside the mind and the possibility of the mind's attaining these things and constructing within itself and by its own activity, beginning with the senses, a knowledge which is true or in conformity with what is.

He argued that bourgeois liberalism and Marxist communism ended up in the same place: treating the human not as person but as an individual and hence as an object, resulting in disastrous consequences for people and society. His humanism was firmly based on his adherence to Thomism, which is interesting given the conventional wisdom that St. Thomas was preoccupied with theistic questions and he rejected the secular humanisms of the early twentieth century, especially fascism and communism, which he argued were de-humanising secular religions. Maritain had a wide influence which extended, some propose, to Martin Luther King who presented an undergraduate paper on his work (King, 1951). While Maritain was seen as one of the leading "Catholic" philosophers of the twentieth century, he was adamant that while philosophy informs theology, it is a separate discipline with boundaries that he was careful not to violate. Furthermore, to disregard his contribution to the debate on these ground would be akin to removing Adam Smith's voice from economics due to his apologia for Presbyterianism in *The Wealth of Nations* (Smith, 1796).

Now I will present some aspect of the common good as outlined in Maritain's major thesis on the subject (2002):

- the personality of the human person tends by nature to communion and has by nature an inner urge to the communication of knowledge (p 47).
- to reach a certain degree of elevation in knowledge people need an education and the help of others (p 48).
- to avoid the extremes of communism and totalitarianism; this good is common to both the whole and the parts into which it flows back and which, in turn, must benefit from it (p 51).
- the common good is not just a system of advantages and utilities but a good in itself or the Ancients expressed it, a *bonum honestum* (p 53).
- it is in the nature of things that people as part of society, should be ordained to the common good and the common work for which the members of the city are assembled (p.65).
- People find themselves by subordinating themselves to the group and the group attains its goal only by serving people (p 66)

Suggestions for Discussion and Further Study

Describe some practical implications of the philosophy of the common good for design and innovation studies.

Is Thomism relevant for present-day philosophical debates on design and innovation? Discuss.

8.4.6 The Wizard-of-Oz

Context *This final vignette borrows an analogy from film and literature to confront a matter of concern for many people in the "digital age". Responsibility, debated in Chap. 3, is very difficult to apportion if the protagonist hides behind a digital screen.*

The Wizard-of-Oz concept originates from the 1899 book by Frank Baum and the 1939 MGM film where Dorothy, the Tinman, the Lion and the Scarecrow follow the yellow brick road to Oz in order to make their requests to the awesome Wizard. In the end, the awe-inspiring ruler of Oz turns out to be just a simulation controlled by a very ordinary human (Biberman, Whitty, & Robins, 1999). In a design and innovation Wizard-of-Oz experiment, the function of the Wizard is played by a human, but the user believes it is a computer. For example, Wizard-of-Oz experiments provide an effective early prototype environment to test interactions between humans and voice recognition applications.

In a debate within the corporate ethics literature, Kerlin (1997) uses the Wizard-of-Oz story to develop an argument that challenges the prevailing wisdom that a group or organisation can be treated as a moral agent per se. His metaphysical thesis is that business ethics is the responsibility of the people that make the decisions and not some abstract corporate entity. These people hide behind the corporate entity in the fashion of the Wizard-of-Oz. Such philosophical reflection has surely become relevant in light of recent high-profile business scandals and could instigate debate on the rationale for differentiation between individual and organisational innovations. In terms of the broader research and educational agenda, the wonderful Wizard-of-Oz has become a popular teaching tool in the area of economics, albeit not without some concerns on the danger of reading too much into the story (Hansen, 2002).

Furthermore, Kerlin argues that the first person facilitates lively writing and avoids any "hesitation to state positions" (Klein, 2005). Kerlin (1997 p. 1437) took issue with Peter French on the metaphysical and practical implications of the latter's work on corporate ethics. Kerlin brings French to task for treating a Corporation as a "moral agent in its own right" since "we cannot reason with the organisation or shame it" and we are unable to attribute *responsibility* to an abstraction. He furthermore emphasises that "our moral discourse is with the creators of the structures" (p. 1437). In relation to this, Willis (2008 p. 355) argues that there is little doubt that the "opinions, theories, biases, and beliefs of researchers have an impact on results. One way to acknowledge that is to use 'I' and 'we' instead of 'the researcher' or 'the researchers' in your papers. Another is to make your background, beliefs, theories, and biases as clear to readers as you can".

Suggestions for Discussion and Further Study

Do present-day technologists accept responsibility for the technology they work on? Discuss.

Describe what can be put in place in the SAL framework to guard against a Wizard-of-Oz syndrome.

8.5 Conclusions

The aim of this chapter was to respond to the editors of this academic series who challenged contributors to stimulate debate and push boundaries in technology education. The chapter was structured in the form of vignettes that provide a short piece written on a subject relevant to the theme of teaching design and innovation. The choice of the topics was taken from the author's research and experience as a practitioner. The subjects are not exhaustive and have consequent limitations. Each vignette is placed in the context of the debates in the previous chapters of the book and concludes with suggestions for discussion and further study. The objective is to provide teachers with topics for dialogue with their students (and colleagues) and to provide material for research. An ancillary aim of the vignettes is to provide teachers and postgraduates with material to develop more detailed bodies of studies in the area.

Chapter Summary

- A research agenda provides an impetus to extend the boundaries of established studies with the aim of further developing a discipline and adding to the body of knowledge.
- Vignettes provide a brief evocative description of a subject (ODE, 2006b).
- The vignettes in this chapter were based on the research and practitioner experience of the author and are a personal reflection on issues within contemporary technology education.
- Each vignette is placed in the context of the debates in the previous chapters of the book.
- Each vignette ends with suggestions for discussion and future study.

References

Bailey, J. I. (2014). Enframing the flesh: Heidegger, transhumanism, and the body as "standing reserve". *Journal of Evolution & Technology, 24*(2), 44–66.

Beadle, R., & Moore, G. (2011). Macintyre, neo-aristotelianism and organization theory. In H. Tsoukas & R. Chia (Eds.), *Philosophy and organization theory research in the sociology of organizations* (Vol. 32, pp. 85–121). Emerald Publishing Limited. ProQuest Ebook Central, http://ebookcentral.proquest.com/lib/gmit/detail.action?docID=662385. Created from GMIT on 2018-12-19 08:24:44.

Bell, S., & Morse, S. (1999). *Sustainability indicators: Measuring the immeasurable?* London/Sterling, VA: Earthscan.

Biberman, J., Whitty, M., & Robins, L. (1999). Lessons from Oz: Balance and wholeness in organisations. *Journal of Organizational Change, 12*(3), 243–254.

Bogen, J. (2005). The Vienna Circle (Wiener Kreis). In T. Honderich (Ed.), *The Oxford companion to philosophy* (2nd ed.). Oxford, UK: Oxford University Press.

Borko, H., Whitcomb, J., & Liston, D. (2008). Wicked problems and other thoughts on issues of technology and teacher learning. *Journal of Teacher Education, 60*(1), 3–7. https://doi.org/10.1177/0022487108328488

Brittan, G. G. J. (1999). Enlightenment. In R. Audi (Ed.), *The Cambridge dictionary of philosophy* (p. 266). New York: Cambridge University Press.

Chesbrough, H. W. (2003). *Open innovation: the new imperative for creating and profiting from technology* Boston: Harvard Business School.

Costello, G. J., & Donnellan, B. (2007). The diffusion of WOZ: Expanding the topology of IS innovations. *Journal of Information Technology, 22*(1), 79–86. https://doi.org/10.1057/palgrave.jit.2000085

Critchley, S. (2001). *Continental philosophy: a very short introduction.* Oxford: Oxford University Press.

Dahlin, B. (2012). Our posthuman futures and education: Homo Zappiens, Cyborgs, and the New Adam. *Futures, 44*(1), 55–63.

Daintith, J., & Gjertsen, D. (Eds.). (1999). *Oxford dictionary of scientists* (2nd ed.). Oxford, UK: Oxford University Press.

Davis, M. (2017). *Teaching design: A guide to curriculum and pedagogy for college design faculty and teachers who use design in their classrooms.* New York: Allworth.

Dimache, A., & Roche, T. (2013). A decision methodology to support servitisation of manufacturing. *International Journal of Operations & Production Management, 33*(11/12), 1435–1457.

DJSI. (2019). Dow Jones sustainability index family available on line through https://www.robecosam.com/csa/indices/?r. Accessed July 2019.

Dodgson, M., Gann, D., & Salter, A. (2005). *Think, play, do: Technology, innovation, and organization.* Oxford, UK: Oxford University Press.

Dreyfus, H., & Dreyfus, S. (1990). Making a mind versus modelling the brain: Artificial intelligence back at the branch point. In M. A. Boden (Ed.), *The philosophy of artificial intelligence (Oxford readings in philosophy)* ((July 12, 1990) reprinted 2005, 1st ed.). Oxford, UK: Oxford University Press.

Dreyfus, H. L. (2007). Why Heideggerian AI failed and how fixing it would require making it more Heideggerian. *Artificial Intelligence, 171*(18), 1137–1160.

EC. (2018). Statement on artificial intelligence, robotics and ‘autonomous’ systems by the European Group on ethics in science and new technologies: European Commission.

Ekvall, G. (1996). Organizational climate for creativity and innovation. *European Journal of Work & Organizational Psychology, 5*(1), 105–123.

Forlano, L. (2017). Posthumanism and design. *She Ji: The Journal of Design, Economics, and Innovation, 3*(1), 16–29.

Fotion, N. (2005). Logical positivism. In T. Honderich (Ed.), *The Oxford companion to philosophy* (2nd ed.). Oxford, UK: Oxford University Press.

Galbraith, J. K. (1967). *The new industrial state* (2nd ed.). London: Pelican Books.

Gregory, B. S. (2012). *The unintended reformation: How a religious revolution secularized society.* Harvard University Press, Cambridge, Massachusetts.

Guignon, C. B. (1999). Heidegger, Martin. In R. Audi (Ed.), *The Cambridge dictionary of philosophy* (p. 610). New York: Cambridge University Press.

Hansen, B. A. (2002). The fable of the allegory: The wizard of Oz in economics. *Journal of Economic Education, 33*(3), 254–264.

Heidegger, M. (1977). The Question Concerning Technology *The Question Concerning Technology: And Other Essays* (pp. 3-35). New York & London Garland Publishing, Inc.

Hirschheim, R. A., Klein, H.-K., & Lyytinen, K. (1995). *Information systems development and data modeling : conceptual and philosophical foundations*. Cambridge: Cambridge University Press.

Hursthouse, R., & Pettigrove, G. (2018). Virtue ethics, In E. N. Zalta (ed.), *The Stanford Encyclopedia of philosophy* (Winter 2018 Edition), Retrieved from https://plato.stanford.edu/archives/win2018/entries/ethics-virtue/.

Inwood, M. J. (2005a). Enlightenment. In T. Honderich (Ed.), *The Oxford companion to philosophy* (2nd ed., pp. 252–253). Oxford, UK: Oxford University Press.

Inwood, M. J. (2005b). Heidegger, Martin. In T. Honderich (Ed.), *The Oxford companion to philosophy* (2nd ed., pp. 371–375). Oxford University Press, Oxford.

Jrade, A., & Jalaei, F. (2013). Integrating building information modelling with sustainability to design building projects at the conceptual stage. *Building Simulation, 6*(4), 429–444.

Kemmis, S., & Smith, T. (2008). Personal praxis: Learning through experience. In P. S. P. Salo & S. Kemmis (Eds.), *Enabling praxis: Challenges for education* (2nd ed., pp. 15–36). Rotterdam, The Netherlands: Sense Publishers.

Kenny, A. (2010). *A new history of western philosophy*. Oxford: Oxford University Press.

Kerlin, M. J. (1997). Peter French, corporate ethics and the wizard of Oz. *Journal of Business Ethics, 16*, 1431–1438.

King, M. L. (1951). Jacques Maritain. In *The papers of Martin Luther King, Jr. (available on-line through* https://kinginstitute.stanford.edu/king-papers/documents/jacques-maritain. Accessed Dec 2019: Stanford: The Martin Luther King, Jr. Research and Education Institute.

Klein, M. (2005). Responsibility. In T. Honderich (Ed.), *The Oxford companion to philosophy* (2nd ed., pp. 815–816). Oxford, UK: Oxford University Press.

Kline, S. J. (2002). What is technology? In R. Scharff & V. Dusek (Eds.), *Philosophy of technology: The technological condition*. Oxford: Blackwell.

Kraut, R. (2018). Aristotle's ethics. Available online through https://plato.stanford.edu/archives/sum2018/entries/aristotle-ethics/. In E. N. Zalta (Ed.), *The Stanford Encyclopedia of philosophy* (Summer 2018 Edition).

Krumdieck, S. (2020). *Transition engineering : Building a sustainable future*. Florida: CRC Press.

Lacy, A. (2005). Positivism. In T. Honderich (Ed.), *The Oxford companion to philosophy* (2nd ed.). Oxford, UK: Oxford University Press.

Laudon, K. C., & Laudon, J. P. (2002). *Management information systems: Managing the digital firm* (2nd ed.). Upper Saddle River, NJ: Prentice Hall.

Macquarrie, J. (1968). *Martin Heidegger*. Richmond, Virginia: John Knox Press.

MacIntyre, A. C. (2002). *A short history of ethics : A history of moral philosophy from the Homeric age to the 20th century*. London: Taylor & Francis Ltd..

MacIntyre, A. C. (2007). *After Virtue* (3rd ed.). University of Notre Dame Press, Notre Dame, Indiana.

Maritain, J. (1932). *An introduction to philosophy (translated by E.I. Watkin)*. London: Sheed & Ward.

Maritain, J. (2002). *The person and the common good (translated by John J. Fitzgerald)*. University of Notre Dame Press; New Impression edition (October 31, 1973).

Moran, D. (2000). *Introduction to phenomenology*. London/New York: Routledge.

Murphy, M. C. (Ed.). (2003). *Alasdair MacIntyre*. Cambridge University Press, Cambridge, United Kingdom.

Murphy, J. B. (1999). Common good. In R. Audi (Ed.), *The Cambridge dictionary of philosophy* (p. 915). New York: Cambridge University Press.

O'Brien, J. A., & Marakas, G. M. (2009). *Management information systems*. Boston: McGraw-Hill Irwin.

ODE. (2006a). technology. In *Oxford dictionary of english* (Revised, 2 ed.). Oxford, UK: Oxford University Press.

ODE. (2006b). vignette. In *Oxford dictionary of english* (Revised, 2 ed.). Oxford University Press.

Pfaffenberger, B. (2002). *Computers in your future*. Upper Saddle River, NJ: Prentice Hall.

Post, G. V., & Anderson, D. L. (2003). *Management information systems : Solving business problems with information technology*. Boston/London: McGraw-Hill/Irwin.

Potapova, A., & Rodionov, S. (2014). Universal empathy and ethical bias for artificial general intelligence. *Journal of Experimental & Theoretical Artificial Intelligence, 26*(3), 405–416.

Pryshlakivsky, J., & Searcy, C. (2013). Sustainable Development as a Wicked Problem In S. F. Kovacic & A. Sousa-Poza (Eds.), *Managing and Engineering in Complex Situations* (pp. 109-128). Dordrecht: Springer Science & Business Media.

Pullen, A., de Weerd-Nederhof, P. C., Groen, A. J., & Fisscher, O. A. M. (2012). Open innovation in practice: Goal complementarity and closed NPD networks to explain differences in innovation performance for SMEs in the medical devices sector. *Journal of Product Innovation Management, 29*(6), 917–934.

Robson, W. (1997). *Strategic management and information systems : An integrated approach* (2nd ed.). London: Pitman Publishing.

Schein, E. H. (1985). *Organizational culture and leadership*. San Francisco: Jossey-Bass Publishers.

Schön, D. A. (1983). *The reflective practitioner: How professionals think in action*. New York: Basic Books.

Scruton, R. (2004). *Modern philosophy: An introduction and survey*. London: Pimlico.

Simon, H. (1996). *The sciences of the artificial* (3rd ed.). The MIT Press, Cambridge, Massachusetts.

Skaburskis, A. (2008). The Origin of "Wicked Problems". *Planning Theory & Practice, 9*(2), 277–280. https://doi.org/10.1080/14649350802041654.

Slappendel, C. (1996). Perspectives on innovation in organizations. *Organization Studies, 17*(1), 107–129.

Slife, B. D., & Richardson, F. C. (2011). The relativism of social constructionism. *Journal of Constructivist Psychology, 24*(4), 333–339. https://doi.org/10.1080/10720537.2011.593475

Smith, A. (1796). Book V, Chapter I: Of the expences of the sovereign or commonwealth. In E. Cannan (Ed.), *An inquiry into the nature and causes of the wealth of nations*. Available online through http://www.econlib.org/LIBRARY/Smith/smWN.html) (1904. 5 ed.). London: Methuen and, Ltd.

Sweet, W. (2004). Jacques Maritain. In E. N. Zalta (Ed.), *The Stanford encyclopedia of philosophy* (Spring ed.). http://plato.stanford.edu/archives/spr2004/entries/maritain/

Uebel, T. (1999). Vienna Circle. In R. Audi (Ed.), *The Cambridge dictionary of philosophy* (p. 915). New York: Cambridge University Press.

Vanhaverbeke, W., & Cloodt, M. (2006). Open innovation in value networks. In H. Chesbrough, W. Vanhaverbeke, & J. West (Eds.), *Open innovation: Researching a new paradigm* (pp. 258–284). Oxford, UK: Oxford University Press.

Waliaula, A. J. (2013). Positivism. *Salem Press Encyclopedia research starters, EBSCOhost*. Accessed 27 March 2018.

Ward, J., Griffiths, P., & Whitmore, P. (1990). *Strategic planning for information systems*. Wiley, Hoboken, New Jersey.

WBCSD. (2019). World Business Council for Sustainable Development (WBCSD) avialable online through https://www.wbcsd.org/. Accessed July 2019.

Weber, R. (2003a). Editor's comments : The problem of the problem. *MIS Quarterly, 27*(1), iii–ix.

Weber, R. (2003b). Editor's comments : Theoretically speaking. *MIS Quarterly, 27*(3), iii–xii.

Weber, R. (2004). The rhetoric of positivism versus interpretivism: A personal view. *MIS Quarterly, 28*(1), iii–xiii.

Wernick, A. (2001). *Auguste Comte and the religion of humanity: The post-theistic program of French Social Theory*. Cambridge: Cambridge University Press, eBook Collection (EBSCOhost), EBSCOhost, viewed 27 March 2018.

West, J., Vanhaverbeke, W., & Chesbrough, H. (2006). Open innovation: A research agenda. In H. Chesbrough, W. Vanhaverbeke, & J. West (Eds.), *Open innovation: Researching a new paradigm* (pp. 258–284). Oxford, UK: Oxford University Press.

Willis, J. (2008). *Qualitative research methods in education and educational technology* Charlotte, NC: Information Age Publishing, eBook Collection (EBSCOhost), EBSCOhost, viewed 22 March 2018.

Wilson, J. Q. (1966). Innovation in organizations: Notes toward a theory. In J. D. Thompson (Ed.), *Approaches to organizational design* (pp. 193–218). Pittsburgh, PA: University of Pittsburgh Press.

Yearworth, M. (2016). Sustainability as a 'super-wicked' problem; opportunities and limits for engineering methodology. *Intelligent Buildings International, 8*(1), 37–47. https://doi.org/10.1080/17508975.2015.1109789

Zaltman, G., Duncan, R., & Holbek, J. (1973). *Innovations and organizations*. New York: Wiley.

MIX
Papier aus verantwortungsvollen Quellen
Paper from responsible sources
FSC® C105338

If you have any concerns about our products,
you can contact us on
ProductSafety@springernature.com

In case Publisher is established outside the EU,
the EU authorized representative is:
Springer Nature Customer Service Center GmbH
Europaplatz 3, 69115 Heidelberg, Germany

Printed by Libri Plureos GmbH
in Hamburg, Germany